Motion Control of Functionally Related Systems

Automation and Control Engineering

Series Editors - Frank L. Lewis, Shuzhi Sam Ge, and Stjepan Bogdan

Synchronization and Control of Multiagent Systems
Dong Sun

System Modeling and Control with Resource-Oriented Petri Nets
MengChu Zhou, Naiqi Wu

Deterministic Learning Theory for Identification, Recognition, and Control
Cong Wang and David J. Hill

Optimal and Robust Scheduling for Networked Control Systems
Stefano Longo, Tingli Su, Guido Herrmann, and Phil Barber

Electric and Plug-in Hybrid Vehicle Networks
Optimization and Control
Emanuele Crisostomi, Robert Shorten, Sonja Stüdli, and Fabian Wirth

Adaptive and Fault-Tolerant Control of Underactuated Nonlinear Systems
Jiangshuai Huang, Yong-Duan Song

Discrete-Time Recurrent Neural Control
Analysis and Application
Edgar N. Sánchez

Control of Nonlinear Systems via PI, PD and PID
Stability and Performance
Yong-Duan Song

Multi-Agent Systems
Platoon Control and Non-Fragile Quantized Consensus
Xiang-Gui Guo, Jian-Liang Wang, Fang Liao, Rodney Swee Huat Teo

Classical Feedback Control with Nonlinear Multi-Loop Systems
With MATLAB® and Simulink®, Third Edition
Boris J. Lurie, Paul Enright

Motion Control of Functionally Related Systems
Tarik Uzunović and Asif Šabanović

For more information about this series, please visit: https://www.crcpress.com/Automation-and-Control-Engineering/book-series/CRCAUTCONENG

Motion Control of Functionally Related Systems

Tarik Uzunović
Faculty of Electrical Engineering, University of
Sarajevo, Bosnia and Herzegovina

Asif Šabanović
Academy of Sciences and Arts of Bosnia and
Herzegovina, and International University of
Sarajevo, Bosnia and Herzegovina

CRC Press
Taylor & Francis Group
Boca Raton London New York

CRC Press is an imprint of the
Taylor & Francis Group, an **informa** business

MATLAB® and Simulink® are trademarks of The MathWorks, Inc. and are used with permission. The MathWorks does not warrant the accuracy of the text or exercises in this book. This book's use or discussion of MATLAB® and Simulink® software or related products does not constitute endorsement or sponsorship by The MathWorks of a particular pedagogical approach or particular use of the MATLAB® and Simulink® software.

CRC Press
Taylor & Francis Group
6000 Broken Sound Parkway NW, Suite 300
Boca Raton, FL 33487-2742

© 2020 by Taylor & Francis Group, LLC
CRC Press is an imprint of Taylor & Francis Group, an Informa business

No claim to original U.S. Government works

Printed on acid-free paper

International Standard Book Number-13: 978-0-367-20880-6 (Hardback)

Visit the Taylor & Francis Web site at
http://www.taylorandfrancis.com

and the CRC Press Web site at
http://www.crcpress.com

We dedicate this book to our families.

Contents

Preface

This book is concerned with the development of design techniques for controlling the motion of mechanical systems which are employed to execute certain tasks acting together collaboratively. When a complex task has to be performed by multiple systems, it imposes functional dependencies between the states and outputs of the systems. These functional dependencies create a system of 'virtually' interconnected subsystems, even though they may be physically separated. The component subsystems within the overall system we call 'functionally related systems', since the nature of the task of the system is defining functional relations between the system components.

The concept of functionality appeared in motion control in earlier works of Toshiaki Tsuji, and it was based on simple motion components that a system can exhibit which are denoted as functions. The concept is further extended in Asif's earlier book on motion control systems where so-called 'functionally related systems' were first defined. After Tarik started his PhD studies under Asif's supervision, we discussed the concept of functionally related systems on a daily basis and came to the conclusion that this concept covers a variety of tasks appearing in motion control. Our intention was to come up with a systematic unified method for control synthesis for these tasks, and demonstrate it experimentally, as well as in simulations.

The intention of this book is to introduce unified control design procedure for functionally related systems, and to show that controllers for different tasks can be successfully created using the procedure. The book starts with an overview of the control methods appearing in the motion control area. Tasks can generally be divided into their components, denoted as functions in the book. It is shown how the dynamics of those tasks can be described. Based on the presented description, several control methods are discussed. The applicability of the introduced control design approach is demonstrated in subsequent chapters for various tasks.

We would like to thank our students whose curiosity inspired us to write this material. We would like to express our sincerest gratitude to our families for their support during years of research and teaching, and particularly during the preparation of the book.

Tarik Uzunović
Asif Šabanović

MATLAB® and Simulink® are registered trademarks of The MathWorks, Inc. For product information, please contact:

The MathWorks, Inc.
3 Apple Hill Drive
Natick, MA, 01760-2098 USA
Tel: 508-647-7000
Fax: 508-647-7001
E-mail: info@mathworks.com
Web: www.mathworks.com

Authors

Tarik Uzunović received the B.Eng. and M.Eng. degrees in electrical engineering from the University of Sarajevo, Sarajevo, Bosnia and Herzegovina, and the Ph.D. degree in mechatronics from Sabanci University, Istanbul, Turkey, in 2008, 2010, and 2015, respectively.

He is an Assistant Professor with the Department of Automatic Control and Electronics, Faculty of Electrical Engineering, University of Sarajevo, Sarajevo, Bosnia and Herzegovina. His research interests include motion control, robotics, and mechatronics.

Asif Šabanović, Emeritus Professor and a member of the Academy of Sciences and Arts of Bosnia and Herzegovina, received the B.S., M.S., and Dr.Sci. degrees in electrical engineering from the University of Sarajevo, Sarajevo, Bosnia and Herzegovina, in 1970, 1975, and 1979, respectively.

From 1970 until 1991, he was with Energoinvest - Institute for Control and Computer Sciences, Sarajevo. In 1991, he was with the Department of Electrical Engineering, University of Sarajevo. He was a Visiting Researcher with the Institute of Control Sciences, Moscow, Russia, Visiting Professor with the California Institute of Technology - CALTECH, Pasadena, Visiting Professor with Keio University, Yokohama, Japan, Full Professor with Yamaguchi University, Ube, Japan, Head of CAD/CAM and the Robotics Department at Tubitak - Marmara Research Centre, Istanbul.

Symbols

Symbol Description

q	configuration vector	e	tracking error vector
v	configuration space velocity vector	f	function vector
A	kinetic energy matrix (inertia matrix)	φ	vector of φ_i functions
		$\mathbf{J}_f$	function Jacobian
b	vector of Coriolis forces, viscous friction forces, and centripetal forces	$\mathbf{J}_f^{\#}$	right pseudoinverse of $\mathbf{J}_f$
		Ω	control transformation matrix
g	vector of gravity terms	$\mathbf{u}_q$	control acceleration
$\mathbf{T}_{ext}$	vector of external forces		
T	vector of generalized joint forces (control vector, input force vector)	$\mathbf{u}_f$	control vector in the function space
		σ	generalized error

1

Introduction

1.1 Motivation

One can often encounter a task which has to be executed by multiple mechanical systems. Therefore, the systems need to operate cooperatively to realize the specified task. Control system design for such a situation is a very challenging mission. It is clear that the nature of the task creates functional relations between the systems and that imposes functional dependencies between the states and outputs of the systems (in one word, coordinates). It can be considered that these functional dependencies make the systems 'virtually' interconnected, even though they may be physically separated. Thus, the systems can be treated as sub-systems of an overall complex system. The component sub-systems within the overall system are called functionally related systems, since the nature of the task is defining functional relations between them.

The conventional approach in motion control systems implies that a task execution in a system that consists of multiple physical units (sub-systems) is typically based on the generation of references for separate sub-systems and control of those subsystems using some of the known control strategies. The definition of these unit tasks to be executed by the sub-systems can be significantly demanding. This makes control of the overall system very complex, as it is sometimes very complicated to find appropriate references for the sub-systems that have to cooperate in order to accomplish a certain task. There is another important disadvantage with this approach. Basically, the task is being decomposed not according to the functions that form the task, but according to the sub-systems that will execute parts of the task. Problems arising with the conventional approach are becoming even more obvious when some function in the task needs to be executed by several sub-systems synchronously, or when a sub-system has to execute multiple functions in the same time. In that situation, it becomes very demanding to generate a reference for a single sub-system's motion. Thus, some new strategy is needed. The aim of this book is to contribute to proposing control of complex motion tasks based on functions.

The following fact can be noticed; it would be reasonable to describe the system task using the functions that have to be controlled. The functions would establish certain functional relationships between coordinates that describe system units (sub-systems). Then, the task would be executed if those

1

functional relationships track their references, where the references are actually references for the functions to be completed. This function-based description makes it possible to control task components, which are identified as the functions. Of course, at the end it is necessary to generate control signals for the independent system units, but the main goal is to find a systematic way to accomplish that mission. If the goal is achieved, a control system designer would only need to describe the desired task by specifying functions that form the task, and the systematic approach could be applied. Systems interconnected with certain functional relationships formed in order to accomplish certain tasks are called functionally related systems. For these systems, a task can be projected to a new coordinate space, denoted as the function space, in which the control design procedure is easier and more intuitive. After a control strategy is designed, one can move back to the original space and calculate control signals that have to be applied to the systems.

Several examples of functionally related systems involved in different tasks are discussed in the subsequent chapters. The examples include: (i) motion synchronization control, (ii) object manipulation control, (iii) formation control of mobile robots, (iv) control of a walking piezoelectric motor, and (v) control in a bilateral system. To explain the motivation for this book, a short discussion about example (ii) will be given here.

Let us assume that two robotic manipulators have to manipulate an object, i.e., to transport it to a specified destination. In order to perform this task, one needs to control grasping force on the object and motion of the object. First, the robots should grasp the object and then move it to the desired destination. Their motion has to be synchronized during the task, and the grasping force needs to controlled at the same time. The task can be very adequately described specifying two functions that have to be controlled: grasping force and motion. Therefore, execution of the whole task can be based on control of these functions.

From the listed examples, and many similar ones that can be created, one can understand that various motion control systems can be designed in the framework based on functionally related systems. Thus, a step towards generalized treatment of such systems is highly desirable. This would simplify control task specification and control strategy design in the motion control systems that can be described within the framework.

1.2 Objectives of the Book

From the short introduction given in the previous subsection, it can be observed that many different technical systems can be categorized as functionally related systems. Nevertheless, this claim will be strongly illustrated throughout this book. The book intends to offer a new perspective for motion control design for such systems. The basic idea is control strategy synthesis based on functions that have to be executed in the controlled system. *Therefore, the main objective of the book is to give a systematic mathematical procedure for motion control design for the functionally related systems.*

In order to accomplish the main objective, other objectives of this book can be listed as follows;

- Definition of an appropriate form to describe function space dynamics of the system with a task containing multiple functions, and dynamics of the system which has to execute several tasks with different priority, or the system including constraints to be satisfied in combination with tasks to be executed.

- Proposal of a systematic and unified method for control design in the function space. The aim is to obtain a unit control distribution matrix in the function space and enforce desired dynamics for each of the identified functions.

- Investigation of two-layer control in two different forms. In the first form, a high-level controller calculates references for low-level controllers, which enforce tracking of these references. In the second form, low-level compensators are compensating disturbances in the configuration space, which is then making the homework for high-level controller easier, since this controller is enforcing desired dynamics of the controlled functions for the system with compensated disturbances.

- Identification of constraints that exist in the selection of functions. The book will examine constraints that have to be taken into account when the function-based approach in control synthesis is used. The constraints will determine which functions can or cannot be accomplished at the same time in the control system, and what are the criteria that have to be taken into consideration when functions are being defined.

- Validation of the proposed approach for control synthesis for several systems. Successful application to different systems will prove the generality of the proposed approach.

2

Methods in Motion Control

This chapter presents an overview about methods in motion control and control of functionally related systems. The chapter starts with a general overview about control methods used in motion control. Then, several examples that can be found in the literature related to the control of functionally related systems are introduced.

2.1 General Structure of Control Systems

The dynamics of a fully actuated mechanical system having n degrees of freedom (n-DOF) can be described in the configuration space as [54]

$$\mathbf{A}(\mathbf{q})\ddot{\mathbf{q}} + \mathbf{b}(\mathbf{q}, \dot{\mathbf{q}}) + \mathbf{g}(\mathbf{q}) + \mathbf{T}_{ext} = \mathbf{T} \qquad (2.1)$$

where

- $\mathbf{q} \in \mathbb{R}^{n \times 1}$ is the configuration vector,
- $\mathbf{A}(\mathbf{q}) \in \mathbb{R}^{n \times n}$ stands for the symmetric positive definite kinetic energy matrix (sometimes referred to as inertia matrix) which has bounded strictly positive elements $0 < a_{ij}^{-} \leq a_{ij}(\mathbf{q}) \leq a_{ij}^{+}$ (hence $0 < A^{-} \leq \|\mathbf{A}(\mathbf{q})\| \leq A^{+}$, where A^{-}, A^{+} are two known scalars satisfying $0 < A^{-} \leq A^{+}$),
- $\mathbf{b}(\mathbf{q}, \dot{\mathbf{q}}) \in \mathbb{R}^{n \times 1}$ represents the vector of Coriolis forces, viscous friction forces, and centripetal forces, and it is bounded by $\|\mathbf{b}(\mathbf{q}, \dot{\mathbf{q}})\| \leq b^{+}$,
- $\mathbf{g}(\mathbf{q}) \in \mathbb{R}^{n \times 1}$ is the vector of gravity terms bounded by $\|\mathbf{g}(\mathbf{q})\| \leq g^{+}$,
- $\mathbf{T}_{ext} \in \mathbb{R}^{n \times 1}$ represents the vector of external forces acting on the system, and
- $\mathbf{T} \in \mathbb{R}^{n \times 1}$ denotes the vector of generalized joint forces, and it will be usually called control vector or input force vector.

It is important to note that (2.1) written in the given form may also be representing the dynamics of several physically separated systems, which can be treated as subsystems of an augmented system having in total n degrees of freedom.

If one desires to control the system described in (2.1), an appropriate control vector $\mathbf{T}$ needs to be determined. For example, the goal can be that configuration vector $\mathbf{q}$ tracks the reference configuration vector $\mathbf{q}^{ref}$. To achieve this goal, a control system designer has to choose the control vector in the following form

$$\mathbf{T} = \mathbf{A}(\mathbf{q})\ddot{\mathbf{q}}^{des} + \mathbf{b}(\mathbf{q}, \dot{\mathbf{q}}) + \mathbf{g}(\mathbf{q}) + \mathbf{T}_{ext} \tag{2.2}$$

where $\ddot{\mathbf{q}} = \ddot{\mathbf{q}}^{des}$ enforces a desired closed-loop dynamics of the controlled system. Different approaches to the design of control (2.2) differ in how to obtain the defined control vector (input force vector).

The first approach is to define the system disturbance (or just disturbance in shorter form) as $\mathbf{T}_d = \mathbf{b}(\mathbf{q}, \dot{\mathbf{q}}) + \mathbf{g}(\mathbf{q}) + \mathbf{T}_{ext}$, which consists of the forces that are mostly unknown, and then use a disturbance observer to estimate the disturbance and apply it as a part of the control vector. The other part of the control vector is $\mathbf{A}(\mathbf{q})\ddot{\mathbf{q}}^{des}$ and it is selected based on a desired closed-loop behavior of the system. Having the disturbance as a component in the control vector makes the system behave in the ideal case as a nominal system $\mathbf{A}(\mathbf{q})\ddot{\mathbf{q}} = \mathbf{A}(\mathbf{q})\ddot{\mathbf{q}}^{des}$ whose dynamics is known. However, in reality disturbance estimation $\hat{\mathbf{T}}_d$ is not equal to the real disturbance, and disturbance estimation error $\mathbf{T}_d - \hat{\mathbf{T}}_d$ is reflected in the resulting dynamics, which is in reality $\mathbf{A}(\mathbf{q})\ddot{\mathbf{q}} = \mathbf{A}(\mathbf{q})\ddot{\mathbf{q}}^{des} - \left(\mathbf{T}_d - \hat{\mathbf{T}}_d \right)$. With a good disturbance compensation, estimation error can be sufficiently small. Then, only desired acceleration $\ddot{\mathbf{q}}^{des}$ needs to be selected and the second component of the input force vector $\mathbf{A}(\mathbf{q})\ddot{\mathbf{q}}^{des}$ can be calculated. If the inertia matrix is unknown, then the matrix can be represented as $\mathbf{A}(\mathbf{q}) = \mathbf{A}_n(\mathbf{q}) + \Delta\mathbf{A}(\mathbf{q})$, where $\mathbf{A}_n(\mathbf{q})$ is the known nominal value and $\Delta\mathbf{A}(\mathbf{q})$ is the unknown variation of the inertia matrix. Selection of $\mathbf{A}_n(\mathbf{q})$ is free; therefore, if $\mathbf{A}(\mathbf{q})$ is known, then $\mathbf{A}_n(\mathbf{q}) = \mathbf{A}(\mathbf{q})$. It is always useful to utilize all available information about the system dynamics for the disturbance estimation. If the term $\Delta\mathbf{A}(\mathbf{q})\ddot{\mathbf{q}}$ is included as a part of the generalized disturbance and if that generalized disturbance $\mathbf{T}_{dis} = \mathbf{T}_d + \Delta\mathbf{A}(\mathbf{q})\ddot{\mathbf{q}}$ is estimated, it is possible to apply the same approach explained above [48, 54]. Thus, in the first approach for the motion control system design, the control system consists of two components, as shown in Figure 2.1. The first one is a disturbance compensator and the second one is a controller that calculates the product of the desired acceleration and inertia matrix. It has to be stated that the control system designer decides which part of the system dynamics will be included in the disturbance, and which part will belong to the nominal system dynamics. If some of the forces in the system are known, or they can be calculated, it is natural to include them in the nominal system dynamics.

The second possible approach in the control of system (2.1) is to have a controller that calculates total control force (2.2) based on all available inputs, as represented in Figure 2.2. In this case it has a double role – to compensate disturbance and to calculate and apply the product of the desired acceleration and inertia matrix.

Based on the above given discussion, all control systems within the motion control area can be classified in two large groups: (i) control systems that do not include disturbance estimation, (ii) control systems that include

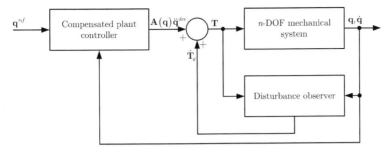

FIGURE 2.1
Control system with disturbance estimation.

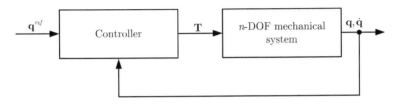

FIGURE 2.2
Control system without disturbance estimation.

disturbance estimation. In the systems that are classified as the systems with disturbance estimation, the control system is a two-loop structure. In the inner loop, uncertainties (structured and/or unstructured) of the controlled system are compensated by a suitable compensator [47, 44, 48]; this was done to obtain a nominal system with known dynamics, considering the controlled mechanical system together with the compensator. That nominal system is then controlled by an outer loop controller whose design was undertaken based on the known nominal system dynamics. Remaining motion control systems, in which control strategy is structured and designed differently, are defined as the systems without disturbance estimation. Both groups include numerous different methods and control algorithms. In the following sections, a survey about both groups is given.

2.1.1 Control Systems without Disturbance Estimation

There are many control systems that do not include estimation of the disturbance. Numerous different algorithms are applied within these systems, and some of them are PID control, sliding mode control, fuzzy control, neural-network-based control, H_∞ control, etc.

As claimed in 2001 by the authors of [3], PID control was the most dominant form of the feedback control used at that time. They also state that PID is used in more than 90% of all then existing control loops. Since PID is a classic and old control method, it is extensively discussed in many control books, for example in [17] and [2]. In motion control applications, PID control has been widely used in a multi-loop structure (inner current loop, middle velocity loop and outer position control), where each loop is controlled by PID (or its derivatives like P, PI) [26]. The literature contains many examples of PID controller applications in motion control systems, and only a few of them will be mentioned here. In [92], a systematic analysis for speed control in a two-inertia system is presented. Straightforward guidelines for PI/PID control system design are given. Three different types of pole assignments are discussed for PI speed control. It is shown that the inertia ratio of load to motor is dominating the damping characteristics of the system. As a significant conclusion, it is stated that different pole-assignment patterns should be considered for different inertia ratios. For an inertia ratio smaller than one, the system becomes seriously underdamped. In that case, PID control can be utilized to improve the two-inertia system performance. Application of P and PID controllers in biaxial contour tracking is discussed in [41]. The proposed control system is based on a coordinate transformation between the X-Y frame and a tangential-contouring (T-C) frame defined along the contour. It is proved that controls of tangential and contour error can be decoupled, if the mismatch between X- and Y-axis dynamics is small. The authors showed the effectiveness of their method with an experiment on a biaxial positioning table, with P and PID controllers used as control laws in tangential and contouring directions. A PID controller is applied as a null space motion controller in the work [58]. In that work, the PID controller is utilized to compensate the disturbance of null space motion without deteriorating asymptotic stability. System performance was better with the PID null space control, when compared to the approach using a null space observer. PID control and its derivatives were extensively used for control of robotic manipulators. Discussions about such applications can be found in [7], [51], [35].

For a few decades, sliding mode control (SMC) has attracted significant interest of researchers and application engineers, mostly due to its simple implementation and good performance. A comprehensive overview of the SMC application in motion control systems is given in [81]. The main feature of the systems with sliding modes is constraint of the system motion in a manifold. Such a motion can appear in systems where control switches between distinct values. Since a different system structure is associated with each of the control values, such systems are called variable structure systems (VSSs). The analysis and design of the systems with sliding modes are discussed in detail in books [14, 28, 70, 71]. A significant portion of the literature on variable structure systems with sliding modes is dedicated to application in electromechanical systems. For example, book [68] is entirely dedicated to that subject. The reader is advised to consider this book as a reference. SMC was very frequently applied for position tracking control of robotic manipulators. The main idea is to construct a manifold in such a way that motion on the manifold implies that the manipulator is tracking the reference position. The first set-point sliding mode controller for robotic manipulators was proposed by Young in 1978 [89]. The control strategy was designed and tested for a two-joint manipulator in a hybrid simulation. As a conclusion, the author stated that the variable structure approach is applicable to the manipulator controller design. This approach was after that often used in position tracking control (and of course in positioning control) for robotic manipulators; discussion and examples of these applications are given in [23, 53]. In addition to the position tracking control, the force control is also discussed in the literature, as an example of SMC application to robotic manipulators, and it is done for both constrained and unconstrained systems. The main idea is the same as for the position tracking, only the manifold description changes. Force control strategies for robotic manipulators with SMC application are discussed in [85, 60, 15, 83]. In the work [69], SMC-based design of the path tracking algorithm for a mobile robot is presented. The authors propose SMC to enforce that a mobile robot tracks a specified gradient. The sliding manifold is designed in such a way that motion on the manifold means that the robot's velocity vector has the specified magnitude and the vector is directed as the specified gradient. A similar approach is given in [21], where SMC is utilized to track the gradient of an artificial potential field. There are many other applications of SMC to motion control systems, with marine vessel control [9], vehicle steering control [6], and control of under-actuated systems [57] being some of them.

A control algorithm based on fuzzy logic [91], that in fact relies on the theory of fuzzy sets [90], is called fuzzy control. A fuzzy controller gives the possibility for mapping of a control strategy described in language to a control algorithm. The control strategy can be based on experience and intuition of a human operator. Thus, the control algorithm is based on expert knowledge. A detailed treatment of fuzzy sets, fuzzy logic and fuzzy control is presented in [8]. Fuzzy controllers have been utilized intensively in motion control, as in many other control fields. In [27], the fuzzy control is applied for indoor navigation and motion control of an autonomous mobile robot (AMR). The author proposes a sensor-based navigation method combining two functions: (i) function for tracing a planned path, (ii) function for avoiding stationary and moving obstacles. Each of these functions is implemented as a separate fuzzy algorithm generating its own control output. The final control output for velocity and steering is a weighted combination of the outputs of two functions, while the weight factor is determined by a separate fuzzy algorithm. In [32], the authors propose a hybrid force/position control of a rehabilitation robot. The hybrid controller is designed to constrain the movement in the desired direction, as well as to maintain a constant force along the moving direction. The authors use a fuzzy position controller and a PI force controller whose parameters are tuned by a fuzzy logic block. Experimental results prove that the robot was able to guide the subjects through planned linear and circular movements. A fuzzy control strategy of roll angle rate and torsion moment for a flexible wing aircraft is presented in [10]. A fuzzy-logic-based system has been utilized to set deflections of the control surfaces on the wings to achieve commanded roll rate, as well as to satisfy prescribed torsion moment constraints. The fuzzy system is described using relatively simple qualitative rules. A hybrid fuzzy control strategy for robotic systems is discussed in [61]. The proposed control system combines a fuzzy gain scheduling method and fuzzy PID controller in order to solve the nonlinear control problem. The basic idea of the proposed method can be described as follows. The fuzzy logic based gain scheduling method is applied to linearize the robotic system at *frozen times*. For each frozen system, a fuzzy PID controller is designed. This controller is created by replacing the PI component of a conventional PID controller with an incremental fuzzy controller.

A neural network (NN) is created as an interconnection of artificial neurons, and it tends to mimic the nervous system of a human brain. Even though one single neuron is very limited in terms of computational power, a huge number of neurons connected in a network significantly increases that power. Applications of NNs in control are discussed in [25]. This technology has been applied in motion control systems in the past. In [82], the authors present an identification and speed control system for a DC motor. An artificial NN is first trained to map inverse dynamics of the motor. That trained network is later combined with a desired reference model to provide speed tracking control. The network is placed in the forward path to cancel the motor dynamics. The suggested method shows how an accurate speed control can be achieved even if the motor and load parameters are unknown. As expected, the parameter variation problem cannot be solved if the off-line training is utilized. NN-based control for a laboratory model helicopter is presented in [79]. The helicopter has two degrees of freedom, the azimuth and elevation angle. NNs applied for control of the two angles are obtained by cloning of previously designed controllers (fuzzy, PID and PID with gain scheduling). Simulation and experimental results are proving that NN-based controllers are providing satisfactory control performance. The control system obtained by cloning fuzzy controllers exhibits the best control performance. A mix locally recurrent NN was used in [11] in order to create a PID-like neural network adaptive controller for uncertain single-input/multi-output systems. The controller consists of an NN that has no more than three nodes in the hidden layer. One activation feedback and an output feedback are included in the hidden layer. The proposed controller can update weights of the NN online, according to control errors, based on stable learning rate. The resilient back-propagation algorithm with a sign instead of the gradient is used to update the network weights. The stability of the closed-loop system is proven, and the effectiveness of the proposed method is demonstrated on single and double inverted pendulums. A robust NN output feedback scheme is created in [37] for the motion control of robotic manipulators. The presented control algorithm does not require measurement of the joint velocities. They are estimated using an NN observer and later used in the controller that is designed as another NN. NN weights are tuned on-line, in both the controller and the observer, with no off-line learning phase needed. This control strategy is applied on a two-link robotic manipulator and given simulation results prove its good performance.

Another strategy utilized in motion control is the H_∞ control method. In this method, a state controller is selected to minimize the H_∞-norm of a closed-loop transfer matrix for linear systems [16] or L_2-gain from the disturbances to the block vector of outputs and inputs for nonlinear systems [80]. Even though this method can provide a robust control law, it was not so extensively used in motion control, when compared to previously discussed methods. In [5], linear robust H_∞ control design methodology is applied to a Space Station attitude and momentum control problem. A robust control synthesis approach is presented for uncertain dynamical systems subject to nonlinear, structured parameter perturbations. The proposed technique incorporates nonlinear, multiparameter variations in the state-space formulation of H_∞ control theory. The authors claim that application of robust H_∞ synthesis technique results in improvement of stability margins with respect to moments-of-inertia uncertainty, compared to conventional linear quadratic regulator design. Nonlinear H_∞ optimal control for agile missiles is discussed in [84]. That method is applied for design of a pitch-plane flight control system for a high-angle-of-attack agile missile. A solution technique for solution of the Hamilton-Jacobi-Isaacs (HJI) partial differential equation is presented, as this solving appears in H_∞ optimal control problems. The approach using successive approximations was applied to a missile flight control problem having six state variables. The authors solved analytically the associated HJI equation for a nonlinear H_∞ inverse-optimal control problem for Euler-Lagrange systems in [49]. As dynamics of the robotic manipulators has this form, the presented result is straightly applicable to robotics. The control based on the given analytic solution consists of reference motion compensation with reference error feedback. By using Lyapunov analysis, the robustness against exogenous disturbance and parametric error was shown. A new technique for controlling the motion of an underactuated vehicle when disturbances are present and only imperfect state measurements are available for feedback is proposed in [65]. The design process starts with a perfect state feedback controller which uses linearization about the desired trajectory to find an optimal control law. The state feedback tracking control law uses an H_∞-optimal design, producing a locally exponentially stable closed-loop system. Later, an H_∞-optimal estimate for the states is included to handle an imperfect state measurement case. In [4], an interesting discussion is given about the implementation of H_∞ control algorithms for sensor constrained mechatronic systems utilizing low cost microcontrollers.

2.1.2 Control Systems with Disturbance Estimation

Control systems categorized as those with disturbance estimation, as already mentioned, have a disturbance compensator whose task is to remove the disturbance and a task controller to enforce desired dynamics of the nominal system. The task controller, designed to enforce the desired dynamics of the nominal system, can belong to any of the groups discussed in the previous section. Thus, in this section several methods for disturbance estimation and its compensation will be considered.

An effective method for obtaining robust control in motion control systems is the utilization of the disturbance observer proposed by Ohnishi in the early 1980s [47, 44, 48]. The disturbance observer is a robust compensator used to estimate the total external force (or torque) acting on a mechanical system, nonlinear gravitational and friction forces, and forces that appear in the system due to parameter variations. The sum of all these forces is mostly denoted as the generalized disturbance. Application of the estimated disturbance as the feedback signal can make the system behave as a nominal system. Therefore, in a motion control system, the generalized disturbance can be fully compensated, since it satisfies the matching condition [13]. When the Ohnishi's disturbance compensator is utilized in the motion control system, the nominal dynamics becomes a double integrator with a known gain (since the nominal inertia is known). Control of such a system is pretty simple and even a basic PD controller with a feedforward term can provide excellent performance. A plant with disturbance observer provides the possibility for control design based on the desired acceleration of the plant. This acceleration can be selected to enforce execution of the task given to the plant. The acceleration control framework was discussed in detail in [54]. The authors showed that the framework can be effectively applied to single-degree of freedom systems, interaction control, bilateral systems, as well as to multibody systems. Disturbance-observer-based control has been used very often in motion control systems. In [48], it is elaborated that the disturbance observer can provide variable stiffness and robustness in a motion control system. That enables its application in position and force control tasks. It is also explained how a disturbance observer can be employed to estimate mechanical parameters of the system. In addition, application of a disturbance observer in motion control of a flexible structure was presented. Control of direct drive motors based on a disturbance observer is discussed in [38]. The authors showed that disturbance can be estimated from the reference current signal and measured velocity or position. Moreover, they have also discussed how poles of the observer can be varied based on command velocity so that precision position control is realized, because the noise of the estimated disturbance influences the position response. Torque sensorless control based on disturbance observer is discussed in [45]. In the proposed method, two disturbance observers are applied to each joint. The first observer is used to realize robust motion controller. The second one is used for obtaining a force sensorless torque controller. The reaction torque is calculated using the estimated disturbance. The effectiveness of the presented method is demonstrated for a multiple degrees of freedom robotic manipulator. The disturbance observer was also utilized in a bilateral system for medical teleoperation [33], human cooperative wheelchair control [34], linear belt drive control [22], contour tracking control for planar stage [18], and trajectory tracking control for planar manipulator [73].

Another method which can provide disturbance compensation in motion control systems is usage of so-called model following control. In this approach, the same reference input is applied to the controlled system as well as to its nominal model. The nominal model is then controlled to track the system output, and a control input is added as additional input to the model. When the tracking is established, the control input is equal to the matched disturbance of the system, and it can be compensated in the controlled system by feeding it back to the system input. This approach was utilized in [1] for disturbance compensation in a piezo stage control. A discrete-time sliding mode controller was used for control of the nominal model. Even open-loop control with compensated disturbance provided high precision. This was further improved by including a closed-loop sliding mode controller in the system. Both methods significantly outperformed classical closed-loop PID control. A similar approach for disturbance compensation was applied in [55]. In a bilateral system, it was necessary to make the slave system behave as a virtual plant. A discrete-time sliding mode controller was utilized to enforce this behavior. Thus, the controller output was in fact removing the disturbance present in the slave system, since this disturbance was not present in the virtual model. Having that the slave system output tracks the virtual model output implies that the present disturbance is compensated. The equivalent approach was additionally used for control of an active four-wheel steering vehicle [24], in order to realize zero sideslip angle and a flat characteristic of frequency response for lateral acceleration at a center of percussion. In this study, a sliding mode controller was applied to make output of the vehicle track output of a linear reference model. Therefore, the control system becomes very robust. The undertaken theoretical analysis confirmed system robustness against system uncertainties such as crosswind disturbance and cornering power perturbations. Moreover, simulation results with a multi degrees-of-freedom vehicle model demonstrated that the proposed control system could make the vehicle run stably even on a slippery road as well as to have fault tolerance against deterioration of the front steering actuator. The model following control technique was applied in motion control systems to make a controlled system behave as a nominal model. Even though there is no explicit disturbance compensation, one can claim that the final result is the same, since the nominal model does not include any uncertainties. Such an example can be found in [88] where model following control for robotic applications is discussed. Again, sliding mode control was used as the control strategy. The presented method was successfully evaluated on a two-link planar manipulator.

The disturbance in a motion control system can as well be compensated using soft computing techniques, namely fuzzy systems and neural networks. Two kinds of adaptive control schemes for a planar robotic manipulator with unstructured and structured uncertainties are presented in [87]. The first scheme is designed to compensate only unstructured uncertainty, while the second one compensates both of these types. In order to compensate the uncertainties, a fuzzy logic system (FLS) is applied, taking into account its capability to approximate a nonlinear function. Robust adaptive control laws are proposed in all proposed schemes for decreasing the effect of approximation error. In that work, the authors have also suggested a method to decrease the total number of fuzzy rules. That can be achieved by considering the properties of robot dynamics and the decomposition of the uncertainty function. An adaptive fuzzy disturbance compensation method is presented in [52], where a control system for a robot is proposed. The adaptation algorithm was derived by using the Lyapunov stability theory. This algorithm provides the global asymptotic stability of the state errors, which results in the sliding-mode regime. Furthermore, the structure of the disturbance compensator is optimized by the introduction of three fuzzy logic subsystems, based on the physical properties of the robot mechanism. A control strategy based on neural networks (NNs) is presented in [40] for the joint-space position control of a mobile manipulator. The arm and the base are separately controlled by two controllers. Each controller output comprises a linear control term (PID) and an NN compensation term. The compensation term is used for on-line estimation of the unknown nonlinear dynamics caused by parameter uncertainty and disturbances. In the suggested method, no preliminary learning stage is required for the NN weights, as on-line learning is utilized. The authors presented a rigorous proof for the tracking stability of the closed-loop system, the convergence of the NN learning process and the boundedness of NN weight estimation errors. Application of NNs as disturbance compensators is further demonstrated in [39], where tracking control in hard disk drives is discussed. In that work, two adaptive NNs are designed for two different tasks. The first NN is designed for disturbance attenuation that appears due to the external vibrations and shocks. The second NN is constructed for friction compensation. The two NN compensators have the appealing advantage that the design does not require any plant, sensor, disturbance, and friction model. The efficiency of the proposed approach is verified through experiments.

2.2   Control of Functionally Related Systems

An idea of functionality is presented in [66, 67]. The authors have identified functions, as the simple motion components that a system can exhibit. Then the system operation can be described as a certain combination of the functions. Depending on the total number of degrees of freedom in the system, a specific number of these functions can be performed simultaneously. A combination of the functions that is being executed in a particular moment defines the current task that the system is executing. It is also shown how certain performance-limit functions, such as the velocity-limit function or position-limit function, can be introduced in the set of functions. The authors state that several standard types of controllers can be used to control functions. These types are listed as rigid coupling controller, spring coupling controller, velocity controller, and force controller. The design approach effectiveness was demonstrated for a parallel link manipulator, used to execute several tasks; in their experiment scenario, one arm of the manipulator was controlled by a human, while others were controlled by a control system whose design was based on different functions that had to be executed for a certain task.

The term 'functionally related systems' was mentioned for the first time in the literature related to motion control systems in [54]. The authors stated in this book that some systems are functionally related if they are 'virtually' interconnected. The term 'virtually' is used to describe a situation in which state or outputs of otherwise separated systems are being functionally related to each other. Several examples of control of functionally related systems are presented, such as bilateral control systems, or systems being controlled to achieve motion synchronization. In each of these systems, in order to execute a system task, the control synthesis can be done as follows. Real state variables are transformed to a virtual space, where the transformation is determined by the imposed task. Afterwards, a control strategy is designed in that virtual space, and virtual control signals are obtained as the result in this phase. In order to be applied to the real physical systems, the virtual signals have to be mapped back to the physical space. However, the discussion about control of functionally related systems given in this book is pretty limited. There is no general procedure that shows a path for control synthesis for those systems. The brief discussion with some illustrative examples shows promising potential of the idea, but it is not fully investigated.

In all mentioned works, the mathematical treatment of control synthesis for functionally related systems is superficial. The idea is described in words, but it is not really mathematically formalized. Moreover, in [66, 67], there is an assumption which could be discussed. The authors use one transformation matrix for mapping the velocities of the motors in the system to the function space velocities, and the same matrix for mapping the motors' accelerations to the function space accelerations. This is correct if and only if the transformation matrix is time independent. Thus, a situation when the matrix is time dependent needs to be discussed. In fact, they have discussed a case when the function space coordinates are expressed as linear functions of the motors' coordinates. A situation when the function space coordinates are expressed as nonlinear functions of the motors' coordinates should as well be considered. It can be concluded that the presented examples of motion control design for functionally related systems in the available literature are not numerous.

This book relies on our recent works [72, 78]. However, our findings from those studies are further generalized and discussed in more detail, especially in terms of control synthesis. Thus, more types of possible tasks in motion control are formally covered.

3

Design of a Motion Control System for Functionally Related Systems

In this chapter, a new approach to design of a motion control system for functionally related systems is presented. The chapter starts with a short discussion about the configuration space dynamics of a mechanical system and synthesis of a control algorithm. After that, it is shown how the dynamics of the system can be represented in the function space, if one task has to be executed by the system. Later, the dynamics of the system is described using the same approach for two tasks with different priority. In addition, it is discussed how the system can be described when constraints and different tasks are present at the same time. Finally, the focus is on the design of a control algorithm which enforces the desired dynamics of constraints and tasks assigned to the mechanical system.

3.1   Configuration Space Control

In this section, a mechanical system with the dynamics described in (2.1) will be discussed. In (2.1), matrix $\mathbf{A}(\mathbf{q})$ is a nonsingular matrix; therefore, this equation can be rewritten as

$$\ddot{\mathbf{q}} + \mathbf{A}^{-1}\left(\mathbf{b} + \mathbf{g} + \mathbf{T}_{ext}\right) = \mathbf{A}^{-1}\mathbf{T} \qquad (3.1)$$

where explicit dependance on the configuration vector and its time derivative is omitted for shorter writing. In this work, the omission of an explicit dependance will often be used. The term $\mathbf{A}^{-1}\mathbf{T}$ is the acceleration in the configuration space introduced by the input force $\mathbf{T}$ with $\mathbf{A}^{-1}$ as the control distribution matrix. This term will be denoted as the control acceleration $\mathbf{u}_q \in \mathbb{R}^{n \times 1}$ in the further derivation, and it is defined as

$$\mathbf{u}_q = \mathbf{A}^{-1}\mathbf{T}. \qquad (3.2)$$

In the configuration space, the control acceleration is considered as a control signal, and input force is taken just as a mean to enforce that acceleration through the inverse inertia matrix as control distribution matrix. From (3.2), it follows

$$\mathbf{T} = \mathbf{A}\mathbf{u}_q. \tag{3.3}$$

Therefore, (3.1) can be written in the form

$$\ddot{\mathbf{q}} + \mathbf{A}^{-1}\left(\mathbf{b} + \mathbf{g} + \mathbf{T}_{ext}\right) = \mathbf{u}_q \tag{3.4}$$

Taking into account that configuration space velocity vector $\mathbf{v}$ is defined as

$$\mathbf{v} = \dot{\mathbf{q}} \tag{3.5}$$

one can write (3.4) as

$$\left.\begin{array}{c} \dot{\mathbf{q}} = \mathbf{v} \\ \dot{\mathbf{v}} + \mathbf{A}^{-1}\left(\mathbf{b} + \mathbf{g} + \mathbf{T}_{ext}\right) = \mathbf{u}_q. \end{array}\right\} \tag{3.6}$$

In this form, the configuration space dynamics has the relative degree equal to one with $\mathbf{u}_q$ as control input, i.e. the order of dynamics in which $\mathbf{u}_q$ appears is equal to one.

A control algorithm in the configuration space will be designed to enforce

$$\ddot{\mathbf{q}} = \ddot{\mathbf{q}}^{des} \Leftrightarrow \dot{\mathbf{v}} = \dot{\mathbf{v}}^{des} \tag{3.7}$$

where $\ddot{\mathbf{q}}^{des} = \dot{\mathbf{v}}^{des} \in \mathbb{R}^{n\times 1}$ is the desired configuration space acceleration [54], selected according to a desired dynamics of the configuration vector. If one considers the second term on the left hand side of (3.4) as the nonlinear disturbance vector $\mathbf{u}_{qdis} \in \mathbb{R}^{n\times 1}$, which is generally unknown, then (3.4) becomes

$$\ddot{\mathbf{q}} + \underbrace{\mathbf{A}^{-1}\left(\mathbf{b} + \mathbf{g} + \mathbf{T}_{ext}\right)}_{\mathbf{u}_{qdis}} = \mathbf{u}_q \Leftrightarrow \ddot{\mathbf{q}} = \mathbf{u}_q - \mathbf{u}_{qdis} \tag{3.8}$$

while (3.6) can be written as

$$\left.\begin{array}{c} \dot{\mathbf{q}} = \mathbf{v} \\ \dot{\mathbf{v}} + \underbrace{\mathbf{A}^{-1}\left(\mathbf{b} + \mathbf{g} + \mathbf{T}_{ext}\right)}_{\mathbf{u}_{qdis}} = \mathbf{u}_q \Leftrightarrow \dot{\mathbf{v}} = \mathbf{u}_q - \mathbf{u}_{qdis}. \end{array}\right\} \tag{3.9}$$

For the sake of unified treatment within this book, the dynamics (3.9) will be dominantly used in the further text.

If $\dot{\mathbf{q}} = \mathbf{v}$ is available (measured or calculated), the disturbance vector can be estimated using the classical disturbance observer for every component of the vector $\mathbf{q}$ [47, 44, 48]. In the disturbance estimation, the gain matrix of the disturbance observer $\mathbf{L} \in \mathbb{R}^{n\times n}$ is a constant diagonal matrix given as

$$\mathbf{L} = \mathrm{diag}\left(l_1, l_2, \ldots, l_n\right), \ l_i > 0, \ i = 1, 2, \ldots, n \tag{3.10}$$

while $\mathbf{z} \in \mathbb{R}^{n\times 1}$ is the intermediate variable in the estimation, selected as a linear combination of the unknown disturbance and known vector $\mathbf{v}$ as

$$\mathbf{z} = \mathbf{u}_{qdis} + \mathbf{L}\mathbf{v}. \tag{3.11}$$

With the disturbance modeled as $\dot{\mathbf{u}}_{qdis} = \mathbf{0}$, the disturbance observer is constructed as

$$\begin{aligned} \dot{\mathbf{z}} &= \mathbf{L}\left(\mathbf{u}_q - \mathbf{z} + \mathbf{L}\mathbf{v}\right) \\ \hat{\mathbf{u}}_{qdis} &= \mathbf{z} - \mathbf{L}\mathbf{v}. \end{aligned} \qquad (3.12)$$

The control goal (3.7) will be achieved if $\mathbf{u}_q$ is calculated as

$$\mathbf{u}_q = \hat{\mathbf{u}}_{qdis} + \dot{\mathbf{v}}^{des}. \qquad (3.13)$$

Thus, the control algorithm is designed in the acceleration control framework with disturbance observer. Considering (3.3), the input force is given by

$$\mathbf{T} = \mathbf{A}\left(\hat{\mathbf{u}}_{qdis} + \dot{\mathbf{v}}^{des}\right). \qquad (3.14)$$

Selection of the desired configuration space acceleration can be done in several ways. For example, for the velocity control, if the configuration space velocity vector is required to track the reference $\mathbf{v}^{ref}\left(t\right) \in \mathbb{R}^{n \times 1}$ that is a differentiable vector-valued function of time, while the asymptotic convergence is wanted, $\dot{\mathbf{v}}^{des}$ can be chosen as

$$\dot{\mathbf{v}}^{des} = \dot{\mathbf{v}}^{ref} - \mathbf{K}_v\left(\mathbf{v} - \mathbf{v}^{ref}\right). \qquad (3.15)$$

In (3.15), $\mathbf{K}_v \in \mathbb{R}^{n \times n}$ is a constant diagonal matrix with positive diagonal entries

$$\mathbf{K}_v = \mathrm{diag}\left(k_{v1}, k_{v2}, \ldots, k_{vn}\right), \ k_{vi} > 0, \ i = 1, 2, \ldots, n. \qquad (3.16)$$

With this control algorithm $\mathbf{v}$ is exponentially converging to $\mathbf{v}^{ref}$ without overshoot, if the perfect disturbance estimation is achieved, i.e., if $\hat{\mathbf{u}}_{qdis} = \mathbf{u}_{qdis}$ holds.

For the position control, the configuration vector needs to track the reference $\mathbf{q}^{ref}\left(t\right) \in \mathbb{R}^{n \times 1}$ that is a two times differentiable vector-valued function of time, and asymptotic convergence can still be desired. In this case, one can select such $\mathbf{v}^{ref}$ so that

$$\mathbf{v} \xrightarrow{t \to \infty} \mathbf{v}^{ref} \Rightarrow \mathbf{q} \xrightarrow{t \to \infty} \mathbf{q}^{ref} \qquad (3.17)$$

Such $\mathbf{v}^{ref}$ can be taken as

$$\mathbf{v}^{ref} = \dot{\mathbf{q}}^{ref} - \mathbf{K}_p\left(\mathbf{q} - \mathbf{q}^{ref}\right) \qquad (3.18)$$

where $\mathbf{K}_p \in \mathbb{R}^{n \times n}$ is a constant diagonal matrix with positive diagonal entries

$$\mathbf{K}_p = \mathrm{diag}\left(k_{p1}, k_{p2}, \ldots, k_{pn}\right), \ k_{pi} > 0, \ i = 1, 2, \ldots, n. \qquad (3.19)$$

The desired acceleration can then again be selected as in (3.15). With this control strategy $\mathbf{q}$ is exponentially converging to $\mathbf{q}^{ref}$ without overshoot, when the perfect disturbance estimation is accomplished.

3.2   System Dynamics

3.2.1   Dynamics for Non-Redundant Task

Suppose that a task assigned to the system (3.6) is described by n functions which describe functional relationships between the components of the vectors $\mathbf{q}$, $\mathbf{v}$ and $\dot{\mathbf{v}}$. The task is realized if the functions are tracking their references. In general, the i-th function is defined as

$$\varphi_i = \varphi_i\left(\mathbf{q}, \mathbf{v}, \dot{\mathbf{v}}\right). \tag{3.20}$$

Since the number of functions is equal to the dimension of the input force vector, this task is considered as non-redundant. The function φ_i may not depend on configuration space acceleration, and it will then have the form

$$\varphi_i = \varphi_i\left(\mathbf{q}, \mathbf{v}\right). \tag{3.21}$$

However, φ_i can be a function of the configuration space velocity $\mathbf{v}$ only

$$\varphi_i = \varphi_i\left(\mathbf{v}\right). \tag{3.22}$$

Furthermore, it can be a function of the configuration vector $\mathbf{q}$ only, i.e., it can have the form

$$\varphi_i = \varphi_i\left(\mathbf{q}\right). \tag{3.23}$$

Our desire here is to have a relative degree of functions with configuration space acceleration as control input equal to one. This will allow a similar treatment to control design as for the configuration space control. Therefore, for each φ_i we will define a new function f_i which will be defined as follows. For the function φ_i given by (3.20), we will assume that this function is linear in acceleration, so it can be written in the following form

$$\varphi_i = \mathbf{j}_{\varphi_i}\dot{\mathbf{v}} + v_i\left(\mathbf{q}, \mathbf{v}\right) \tag{3.24}$$

where $\mathbf{j}_{\varphi_i} \in \mathbb{R}^{1 \times n}$ is a vector valued function of the configuration vector and configuration space velocity, and $v_i\left(\mathbf{q}, \mathbf{v}\right)$ is a scalar valued function. The function f_i will be defined by the following equation

$$\dot{f}_i = \varphi_i = \mathbf{j}_{\varphi_i}\dot{\mathbf{v}} + v_i\left(\mathbf{q}, \mathbf{v}\right). \tag{3.25}$$

If the function φ_i is of the form (3.21) or (3.22), f_i is given as

$$f_i = \varphi_i \tag{3.26}$$

while for the φ_i given by (3.23), f_i is defined by

$$f_i = \dot{\varphi}_i. \tag{3.27}$$

It is assumed that φ_i in the form (3.21) or (3.22) is differential with respect to time, while φ_i defined by (3.23) is two times differentiable with respect to time.

The first-order dynamics of the function f_i can now be calculated depending on its definition and function φ_i. In the case when φ_i has the form (3.24), the dynamics of f_i is given by (3.25). For φ_i from (3.21) and f_i given by (3.26), the dynamics is

$$\dot{f}_i = \dot{\varphi}_i = \frac{\partial \varphi_i}{\partial \mathbf{v}} \dot{\mathbf{v}} + \frac{\partial \varphi_i}{\partial \mathbf{q}} \dot{\mathbf{q}} = \mathbf{j}_{\varphi_i} \dot{\mathbf{v}} + \frac{\partial \varphi_i}{\partial \mathbf{q}} \mathbf{v}, \ \mathbf{j}_{\varphi_i} = \frac{\partial \varphi_i}{\partial \mathbf{v}}. \tag{3.28}$$

In the case when φ_i is given by (3.22), the dynamics of f_i which, defined by (3.26), becomes

$$\dot{f}_i = \dot{\varphi}_i = \frac{\partial \varphi_i}{\partial \mathbf{v}} \dot{\mathbf{v}} = \mathbf{j}_{\varphi_i} \dot{\mathbf{v}}, \ \mathbf{j}_{\varphi_i} = \frac{\partial \varphi_i}{\partial \mathbf{v}}. \tag{3.29}$$

When φ_i is defined by (3.23), f_i is given in (3.27), and its dynamics is

$$\dot{f}_i = \ddot{\varphi}_i = \mathbf{j}_{\varphi_i} \dot{\mathbf{v}} + \dot{\mathbf{j}}_{\varphi_i} \mathbf{v}, \ \mathbf{j}_{\varphi_i} = \frac{\partial \varphi_i}{\partial \mathbf{q}}. \tag{3.30}$$

One can now write dynamics of f_i in the general form

$$\dot{f}_i = \mathbf{j}_{\varphi_i} \dot{\mathbf{v}} + v_i(\mathbf{q}, \mathbf{v}) \tag{3.31}$$

which is equivalent to (3.25) when φ_i is given by (3.24). In addition, $v_i(\mathbf{q}, \mathbf{v}) = \frac{\partial \varphi_i}{\partial \mathbf{q}} \mathbf{v}$ for φ_i defined as in (3.21), $v_i(\mathbf{q}, \mathbf{v}) = 0$ for φ_i in the form (3.22), and $v_i(\mathbf{q}, \mathbf{v}) = \dot{\mathbf{j}}_{\varphi_i} \mathbf{v}$, $\mathbf{j}_{\varphi_i} = \frac{\partial \varphi_i}{\partial \mathbf{q}}$ for φ_i given by (3.23). All n functions f_i can be combined in a single vector which will be denoted as the function vector $\mathbf{f}(\mathbf{q}, \mathbf{v}) = [f_1 \ f_2 \dots f_n]^{\mathrm{T}}$.

Now the task can be reformulated in terms of functions f_i and it can be stated that the task assigned to the system (3.6) is realized if functional relationships between the components of the configuration vector and configuration space velocity vector described by n functions written in a vector form as $\mathbf{f}(\mathbf{q}, \mathbf{v}) \in \mathbb{R}^{n \times 1}$, are tracking their references. The vector $\mathbf{f}$ will be called the function vector, as already mentioned. The components of the function vector are differentiable with respect to time.

Based on (3.31), the first-order dynamics of the function vector is

$$\dot{\mathbf{f}} = \begin{bmatrix} \dot{f}_1 \\ \dot{f}_2 \\ \vdots \\ \dot{f}_n \end{bmatrix} = \begin{bmatrix} \mathbf{j}_{\varphi_1} \\ \mathbf{j}_{\varphi_2} \\ \vdots \\ \mathbf{j}_{\varphi_n} \end{bmatrix} \dot{\mathbf{v}} + \begin{bmatrix} v_1(\mathbf{q}, \mathbf{v}) \\ v_2(\mathbf{q}, \mathbf{v}) \\ \vdots \\ v_n(\mathbf{q}, \mathbf{v}) \end{bmatrix} = \mathbf{J}_f(\mathbf{q}, \mathbf{v}) \dot{\mathbf{v}} + \mathbf{\Upsilon} = \mathbf{J}_f(\mathbf{q}, \mathbf{v}) \ddot{\mathbf{q}} + \mathbf{\Upsilon}. \tag{3.32}$$

It will be assumed that the function Jacobian matrix $\mathbf{J}_f \in \mathbb{R}^{n \times n}$ defined by

$$\mathbf{J}_f = \begin{bmatrix} \mathbf{j}_{\varphi_1} \\ \mathbf{j}_{\varphi_2} \\ \vdots \\ \mathbf{j}_{\varphi_n} \end{bmatrix} \qquad (3.33)$$

is a full rank matrix, i.e., $\text{rank}\,(\mathbf{J}_f) = n$.

With (3.6) taken into account, (3.32) becomes

$$\dot{\mathbf{f}} = \mathbf{J}_f \left[\mathbf{u}_q - \mathbf{A}^{-1} \left(\mathbf{b} + \mathbf{g} + \mathbf{T}_{ext} \right) \right] + \boldsymbol{\Upsilon}. \qquad (3.34)$$

Dynamics (3.34) can also be rewritten as

$$\dot{\mathbf{f}} = \mathbf{J}_f \mathbf{u}_q - \mathbf{J}_f \mathbf{A}^{-1} \left(\mathbf{b} + \mathbf{g} + \mathbf{T}_{ext} \right) + \boldsymbol{\Upsilon}. \qquad (3.35)$$

In order to proceed with control design, it is necessary to identify a control vector in the function space. It will be assumed that the control acceleration in the configuration space $\mathbf{u}_q$ is related to the control vector in the function space $\mathbf{u}_f \in \mathbb{R}^{n \times 1}$ by the following equation

$$\mathbf{u}_q = \boldsymbol{\Omega} \mathbf{u}_f. \qquad (3.36)$$

In (3.36), $\boldsymbol{\Omega} \in \mathbb{R}^{n \times n}$ is a control transformation matrix. The control transformation matrix will be determined in order to achieve specific design goals. The second and third term on the right hand side of (3.35) can be identified as a nonlinear disturbance vector.

It is now required to discuss the possible structure of the transformation matrix $\boldsymbol{\Omega}$. At this point, the design requirement is to have one component of the control vector $\mathbf{u}_f$ directly controlling one and only one component of the vector $\mathbf{f}$. Namely, the control distribution matrix in the function space has to be an identity matrix. This requirement can be expressed as

$$\mathbf{J}_f \boldsymbol{\Omega} = \mathbf{I}. \qquad (3.37)$$

As the function Jacobian matrix is a nonsingular matrix, then $\boldsymbol{\Omega}$ which satisfies (3.37) is given as

$$\boldsymbol{\Omega} = \mathbf{J}_f^{-1}. \qquad (3.38)$$

With $\boldsymbol{\Omega}$ defined in this manner, the control vectors in the configuration space and function space are related with

$$\mathbf{u}_q = \mathbf{J}_f^{-1} \mathbf{u}_f \qquad (3.39)$$

which may alternatively be written as

$$\mathbf{u}_f = \mathbf{J}_f \mathbf{u}_q. \qquad (3.40)$$

Taking (3.40) into account, (3.35) becomes

$$\dot{\mathbf{f}} = \mathbf{u}_f - \mathbf{J}_f \mathbf{A}^{-1} \left(\mathbf{b} + \mathbf{g} + \mathbf{T}_{ext} \right) + \boldsymbol{\Upsilon}. \qquad (3.41)$$

The last equation represents the dynamics of the system in the function space.

3.2.2   Conventional and Proposed Method for Control Mapping

In order to investigate the proposed method for control mapping, let us now consider a case in which all functions φ_i are functions of the configuration vector $\mathbf{q}$ only, i.e., they have the form (3.23). In that case, one can calculate functions f_i as

$$
\begin{bmatrix} \dot{\varphi}_1 \\ \dot{\varphi}_2 \\ \vdots \\ \dot{\varphi}_n \end{bmatrix} = \begin{bmatrix} f_1 \\ f_2 \\ \vdots \\ f_n \end{bmatrix} = \begin{bmatrix} \frac{\partial \varphi_1}{\partial \mathbf{q}} \\ \frac{\partial \varphi_2}{\partial \mathbf{q}} \\ \vdots \\ \frac{\partial \varphi_n}{\partial \mathbf{q}} \end{bmatrix} \dot{\mathbf{q}} = \mathbf{J}_f \dot{\mathbf{q}}.
\tag{3.42}
$$

From (3.40), it can be concluded that control signals from the configuration space were mapped to the function space with the same transformation matrix used for mapping velocities from configuration space to the function space [compare (3.42) and (3.40)].

It is now useful to compare (3.39) with the conventionally used mapping of control signals from the task space (corresponding to the function space in this discussion) to the configuration space. In the conventional approach, it is typical to execute this mapping using the transposed function Jacobian matrix. In that form, it would be

$$
\mathbf{u}_q = \mathbf{J}_f^{\mathrm{T}} \mathbf{u}_f^c
\tag{3.43}
$$

where the superscript c represents the control vector for conventional transformation method. The matrix inversion and matrix transpose operations have important properties according to which for any nonsingular matrix $\mathbf{S}$ the following is valid: $\left(\mathbf{S}^{\mathrm{T}}\right)^{-1} = \left(\mathbf{S}^{-1}\right)^{\mathrm{T}} = \mathbf{S}^{-\mathrm{T}}$. If the function Jacobian matrix is a nonsingular matrix, as it is assumed earlier, one can write

$$
\mathbf{u}_f^c = \mathbf{J}_f^{-\mathrm{T}} \mathbf{u}_q.
\tag{3.44}
$$

The last equation implies that mapping of the control signals from the configuration space to the function space, is done with the $\mathbf{J}_f^{-\mathrm{T}}$ transformation matrix. On the other hand, it was proposed in this book that the same mapping should be executed with the $\mathbf{J}_f$ transformation matrix. It will now be analyzed how these two transformations differ, with a different transformation matrix being used.

It is now assumed that the function Jacobian matrix is written in the form

$$
\mathbf{J}_f = \begin{bmatrix} \mathbf{r}_1 \\ \mathbf{r}_2 \\ \vdots \\ \mathbf{r}_n \end{bmatrix}
\tag{3.45}
$$

where each $\mathbf{r}_i$ $(i = 1, 2, \ldots, n)$ is an n-dimensional vector row. Mapping described by (3.40) indicates that k-th component of the vector $\mathbf{u}_f$ is obtained as the scalar product of the vector $\mathbf{u}_q$ and $\mathbf{r}_k$ $(k = 1, 2, \ldots, n)$. If that component is denoted as u_{fk}, it can be written

$$u_{fk} = \mathbf{r}_k \cdot \mathbf{u}_q. \tag{3.46}$$

Considering the definition of the scalar product between vectors, the proposed mapping (3.40) can be interpreted in such a way that components of the function space control vector are obtained as scaled projections of the vector $\mathbf{u}_q$ to the rows of the function Jacobian matrix. Since the function Jacobian matrix is nonsingular, its rows form a basis for n-dimensional space, and control mapping is actually projection of a vector to that basis. As the same mapping is employed for velocities, then the control vector and velocity vector from the configuration space are projected to the same coordinate frame to obtain the control vector in the function space and function vector.

The matrix $\mathbf{J}_f^{-\mathrm{T}}$ can be written in the form

$$\mathbf{J}_f^{-\mathrm{T}} = \begin{bmatrix} \mathbf{r}_1^c \\ \mathbf{r}_2^c \\ \vdots \\ \mathbf{r}_n^c \end{bmatrix} \tag{3.47}$$

where each $\mathbf{r}_i^c$ $(i = 1, 2, \ldots, n)$ is an n-dimensional vector row. Mapping (3.44) implies that the k-th component of the vector $\mathbf{u}_f^c$ is obtained as the scalar product of the vector $\mathbf{u}_q$ and $\mathbf{r}_k^c$ $(k = 1, 2, \ldots, n)$. If that component is denoted as u_{fk}^c, it can be written

$$u_{fk}^c = \mathbf{r}_k^c \cdot \mathbf{u}_q. \tag{3.48}$$

The control vector in the function space is now obtained by projecting the control vector from the configuration space to the coordinate frame defined by the rows of the $\mathbf{J}_f^{-\mathrm{T}}$ matrix.

Let us finally discuss the difference between the frame defined by $\mathbf{r}_i$ and the one defined by $\mathbf{r}_i^c$. Using the properties of the matrix inversion and matrix transpose operations stated above, it can be shown that

$$\mathbf{r}_i \cdot \mathbf{r}_j^c = \begin{cases} 1, & i = j \\ 0, & i \neq j. \end{cases} \tag{3.49}$$

This indicates that the i-th component of the base defined by the Jacobian transformation matrix is orthogonal to all components of the base defined by conventional transformation matrix $\mathbf{J}_f^{-\mathrm{T}}$, except the i-th component. The scalar product of the i-th components of the two bases is equal to one. This represents the relation between the two bases. In addition, it means that projection of the vector $\mathbf{r}_i$ to the vector $\mathbf{r}_i^c$ has magnitude equal to $1/\|\mathbf{r}_i^c\|$, and its projection to other vectors $\mathbf{r}_j^c$, $j \neq i$ is zero. For a two-dimensional system, this relation can be illustrated as in Figure 3.1. It is important to mention here that the illustrated transformation was discussed in [56], but from a very different view point.

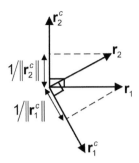

FIGURE 3.1
Illustration of the mutual relation between the frame defined by the function
Jacobian matrix and the frame defined by the inverse transposed function
Jacobian matrix

3.2.3   Dynamics for Redundant Task

In the previous discussion, it was assumed that the function Jacobian matrix
is a nonsingular square matrix. Nevertheless, this does not have to be valid in
all cases. For example, if $\mathbf{f}$ is an m-dimensional vector, with $m < n$, then the
function Jacobian matrix will be $\mathbf{J}_f \in \mathbb{R}^{m \times n}$ and it is still defined as in (3.33).
The first-order dynamics of the function vector is again given in the form
(3.35). The function Jacobian matrix is not a square matrix any more, and $\mathbf{J}_f^{-1}$
does not exist. However, if the control vector in the function space is assumed
to be m-dimensional, it is still possible to achieve the goal that one component
of the control vector $\mathbf{u}_f$ is directly controlling one and only one component of
the vector $\mathbf{f}$. This can be ensured using a right pseudoinverse $\mathbf{J}_f^{\#} \in \mathbb{R}^{n \times m}$ of
the matrix $\mathbf{J}_f$. Then it will be taken $\mathbf{\Omega} = \mathbf{J}_f^{\#}$. Generally, the control of the m-
dimensional function vector requires m independent control inputs, i.e., $\mathbf{u}_f \in
\mathbb{R}^{m \times 1}$. This control vector will be mapped back to the configuration space by
$\mathbf{J}_f^{\#} \mathbf{u}_f$. Still, the redundancy in the configuration space control acceleration
$\mathbf{u}_q$ exists, since $n > m$. Let the component of the configuration space control
acceleration that complements $\mathbf{J}_f^{\#} \mathbf{u}_f$ in order to realize arbitrary acceleration
be expressed as $\mathbf{\Gamma} \mathbf{u}_{q_0}$, where $\mathbf{\Gamma} \in \mathbb{R}^{n \times n}$ is yet to be determined and $\mathbf{u}_{q_0}$ is an
arbitrary acceleration vector in $\mathbb{R}^{n \times 1}$. Then the control acceleration generated
by these two components is

$$\mathbf{u}_q = \mathbf{J}_f^{\#} \mathbf{u}_f + \mathbf{\Gamma} \mathbf{u}_{q_0}. \tag{3.50}$$

When (3.50) is included in (3.35), (3.35) becomes

$$\dot{\mathbf{f}} = \mathbf{u}_f + \mathbf{J}_f \mathbf{\Gamma} \mathbf{u}_{q_0} - \mathbf{J}_f \mathbf{A}^{-1} \left(\mathbf{b} + \mathbf{g} + \mathbf{T}_{ext} \right) + \mathbf{\Upsilon}. \tag{3.51}$$

In order to have $\mathbf{f}$ controlled directly only by $\mathbf{u}_f$, i.e., to achieve decoupling
in terms of control between the function vector and acceleration vector $\mathbf{u}_{q_0}$,

one needs to find $\mathbf{\Gamma}$ such that $\mathbf{J}_f \mathbf{\Gamma} \mathbf{u}_{q_0} = \mathbf{0}$. If $\mathbf{\Gamma}$ is selected as the null space projection matrix of the matrix $\mathbf{J}_f$, then it is

$$\mathbf{\Gamma} = \mathbf{I} - \mathbf{J}_f^{\#} \mathbf{J}_f \tag{3.52}$$

as well as

$$\mathbf{J}_f \mathbf{\Gamma} = \mathbf{J}_f - \underbrace{\mathbf{J}_f \mathbf{J}_f^{\#}}_{\mathbf{I}} \mathbf{J}_f = \mathbf{0}. \tag{3.53}$$

Therefore, the desired decoupling is achieved with $\mathbf{\Gamma}$ given by (3.52). In that case, (3.51) becomes

$$\dot{\mathbf{f}} = \mathbf{u}_f - \mathbf{J}_f \mathbf{A}^{-1} \left(\mathbf{b} + \mathbf{g} + \mathbf{T}_{ext} \right) + \mathbf{\Upsilon}. \tag{3.54}$$

One can notice that (3.54) is written in the identical form as (3.41), but dimensions of some vectors and matrices are different in these two equations. Evidently, the decoupling is achieved with any right pseudoinverse matrix $\mathbf{J}_f^{\#}$. One particular pseudoinverse can be selected according to some criterion. Here, we will discuss two possible situations.

Let us discuss the situation in which all functions φ_i are functions of the configuration vector $\mathbf{q}$ only, and the function vector $\mathbf{f}$ is given by (3.42). Configuration space velocities can be expressed as

$$\dot{\mathbf{q}} = \mathbf{J}_f^{\#} \mathbf{f} + \mathbf{\Gamma} \dot{\mathbf{q}}_0 \tag{3.55}$$

where $\dot{\mathbf{q}}_0$ is any vector in $\mathbb{R}^{n \times 1}$. The pseudoinverse will be selected such that kinetic energy $T = 0.5 \dot{\mathbf{q}}^T \mathbf{A} \dot{\mathbf{q}}$ of the system (2.1) is minimized under the constraint $\mathbf{f} = \mathbf{J}_f \dot{\mathbf{q}}$. In this case, the inertia matrix is used as a weighting matrix, and a weighted pseudoinverse will be obtained as the solution for the pseudoinverse. The problem can be solved by introducing the Lagrange multipliers $\mathbf{\Lambda} \in \mathbb{R}^{m \times 1}$ and forming the modified cost function as

$$\Pi \left(\dot{\mathbf{q}}, \mathbf{\Lambda} \right) = \frac{1}{2} \dot{\mathbf{q}}^T \mathbf{A} \dot{\mathbf{q}} + \mathbf{\Lambda}^T \left(\mathbf{f} - \mathbf{J}_f \dot{\mathbf{q}} \right). \tag{3.56}$$

The solution has to satisfy necessary conditions

$$\left(\frac{\partial \Pi}{\partial \dot{\mathbf{q}}} \right)^T = \mathbf{0} \Rightarrow \mathbf{A} \dot{\mathbf{q}} - \mathbf{J}_f^T \mathbf{\Lambda} = \mathbf{0} \tag{3.57}$$

$$\left(\frac{\partial \Pi}{\partial \mathbf{\Lambda}} \right)^T = \mathbf{0} \Rightarrow \mathbf{f} - \mathbf{J}_f \dot{\mathbf{q}} = \mathbf{0}. \tag{3.58}$$

From (3.57), it follows

$$\dot{\mathbf{q}} = \mathbf{A}^{-1} \mathbf{J}_f^T \mathbf{\Lambda}. \tag{3.59}$$

Using (3.59) in (3.58), it can be written

$$\mathbf{f} = \mathbf{J}_f \dot{\mathbf{q}} = \mathbf{J}_f \mathbf{A}^{-1} \mathbf{J}_f^T \mathbf{\Lambda} \Rightarrow \mathbf{\Lambda} = \left(\mathbf{J}_f \mathbf{A}^{-1} \mathbf{J}_f^T \right)^{-1} \mathbf{f}. \tag{3.60}$$

When (3.60) is used in (3.59), the following can be obtained

$$\dot{\mathbf{q}} = \mathbf{A}^{-1}\mathbf{J}_f^T \left(\mathbf{J}_f \mathbf{A}^{-1}\mathbf{J}_f^T\right)^{-1} \mathbf{f}. \tag{3.61}$$

From the last equation it follows that the pseudoinverse which minimizes kinetic energy is

$$\mathbf{J}_f^{\#} = \mathbf{A}^{-1}\mathbf{J}_f^T \left(\mathbf{J}_f \mathbf{A}^{-1}\mathbf{J}_f^T\right)^{-1}. \tag{3.62}$$

For the generalized inverse from the last equation, Khatib says it is dynamically consistent with the dynamic equations of the controlled robot manipulator and end-effector [36].

In general, all functions φ_i are not functions of the configuration vector $\mathbf{q}$ only. Therefore, the function vector is formed as explained in Section 3.2.1. Its dynamics is given by (3.32). Configuration space accelerations can be expressed as

$$\ddot{\mathbf{q}} = \mathbf{J}_f^{\#}\left(\dot{\mathbf{f}} - \boldsymbol{\Upsilon}\right) + \boldsymbol{\Gamma}\ddot{\mathbf{q}}_0 \tag{3.63}$$

where $\ddot{\mathbf{q}}_0$ is any vector in $\mathbb{R}^{n \times 1}$. The pseudoinverse will be selected such that 'acceleration energy' $T_{\ddot{q}} = 0.5\ddot{\mathbf{q}}^T\mathbf{A}\ddot{\mathbf{q}}$ of the system (2.1) is minimized under the constraint $\dot{\mathbf{f}} = \mathbf{J}_f\ddot{\mathbf{q}} + \boldsymbol{\Upsilon}$. Here, the inertia matrix is again used as a weighting matrix, and a weighted pseudoinverse will be obtained as the solution for the pseudoinverse. The problem is again solved by introducing the Lagrange multipliers $\boldsymbol{\Lambda} \in \mathbb{R}^{m \times 1}$ and forming the modified cost function as

$$\Pi\left(\ddot{\mathbf{q}}, \boldsymbol{\Lambda}\right) = \frac{1}{2}\ddot{\mathbf{q}}^T\mathbf{A}\ddot{\mathbf{q}} + \boldsymbol{\Lambda}^T\left(\dot{\mathbf{f}} - \boldsymbol{\Upsilon} - \mathbf{J}_f\ddot{\mathbf{q}}\right). \tag{3.64}$$

The solution needs to satisfy the following necessary conditions

$$\left(\frac{\partial\Pi}{\partial\ddot{\mathbf{q}}}\right)^T = 0 \Rightarrow \mathbf{A}\ddot{\mathbf{q}} - \mathbf{J}_f^T\boldsymbol{\Lambda} = 0 \tag{3.65}$$

$$\left(\frac{\partial\Pi}{\partial\boldsymbol{\Lambda}}\right)^T = 0 \Rightarrow \dot{\mathbf{f}} - \boldsymbol{\Upsilon} - \mathbf{J}_f\ddot{\mathbf{q}} = 0. \tag{3.66}$$

From (3.65) it follows

$$\ddot{\mathbf{q}} = \mathbf{A}^{-1}\mathbf{J}_f^T\boldsymbol{\Lambda}. \tag{3.67}$$

Using (3.67) in (3.66), one can write

$$\dot{\mathbf{f}} = \mathbf{J}_f\ddot{\mathbf{q}} + \boldsymbol{\Upsilon} = \mathbf{J}_f\mathbf{A}^{-1}\mathbf{J}_f^T\boldsymbol{\Lambda} + \boldsymbol{\Upsilon} \Rightarrow \boldsymbol{\Lambda} = \left(\mathbf{J}_f\mathbf{A}^{-1}\mathbf{J}_f^T\right)^{-1}\left(\dot{\mathbf{f}} - \boldsymbol{\Upsilon}\right). \tag{3.68}$$

When (3.68) is used in (3.67), the following is obtained

$$\ddot{\mathbf{q}} = \mathbf{A}^{-1}\mathbf{J}_f^T\left(\mathbf{J}_f\mathbf{A}^{-1}\mathbf{J}_f^T\right)^{-1}\left(\dot{\mathbf{f}} - \boldsymbol{\Upsilon}\right). \tag{3.69}$$

From the last equation it follows that the pseudoinverse which minimizes 'acceleration energy' is given by

$$\mathbf{J}_f^{\#} = \mathbf{A}^{-1}\mathbf{J}_f^T\left(\mathbf{J}_f\mathbf{A}^{-1}\mathbf{J}_f^T\right)^{-1}. \tag{3.70}$$

Therefore, the matrix $\mathbf{\Omega}$ will be taken as the right pseudoinverse which minimizes the kinetic energy or 'acceleration energy' of the controlled system

$$\mathbf{\Omega} = \mathbf{J}_f^\# = \mathbf{A}^{-1}\mathbf{J}_f^T \left(\mathbf{J}_f \mathbf{A}^{-1}\mathbf{J}_f^T\right)^{-1} \tag{3.71}$$

3.3 Dynamics of System with Tasks in Hierarchical Structure

In this section, it will be assumed that two tasks are supposed to be executed by the system (3.6). The first task is a higher priority task and it is described by m functions $(m < n)$, written in a vector form as $\mathbf{f}_1(\mathbf{q}) \in \mathbb{R}^{m \times 1}$, where the vector $\mathbf{f}_1$ is differentiable with respect to time. It will be assumed that the function Jacobian matrix for the first task $\mathbf{J}_{f_1} \in \mathbb{R}^{m \times n}$ is a full row rank matrix, i.e., $\operatorname{rank}(\mathbf{J}_{f_1}) = m$.

The second, lower priority, task is described by a $(n-m)$-dimensional function vector $\mathbf{f}_2 \in \mathbb{R}^{(n-m) \times 1}$, that is also differentiable with respect to time. As for the previous task, it is assumed that the function Jacobian for the lower priority task $\mathbf{J}_{f_2} \in \mathbb{R}^{(n-m) \times n}$ is a full row rank matrix. Therefore, $\operatorname{rank}(\mathbf{J}_{f_2}) = (n-m)$.

The control goal is to achieve decoupling of the two tasks in terms of control. In other words, it is desired that the control vector associated with one task does not directly control the function vector associated with the other task. In addition, the decoupling should fulfill the aim that a control vector associated with a task is appearing in the dynamics of its corresponding function vector with the unit control distribution matrix. Therefore, it is necessary to find appropriate mapping of the dynamics from the configuration space to the function space in order to achieve the decoupling. Taking this into account, the first-order dynamics of the two function vectors will be written as

$$\begin{bmatrix} \dot{\mathbf{f}}_1 \\ \dot{\mathbf{f}}_2 \end{bmatrix} = \begin{bmatrix} \mathbf{J}_{f_1} \\ \mathbf{J}_{f_2}^* \end{bmatrix} \dot{\mathbf{v}} + \begin{bmatrix} \mathbf{\Upsilon}_1 \\ \mathbf{\Upsilon}_2 \end{bmatrix} = \mathbf{J}_f \dot{\mathbf{v}} + \mathbf{\Upsilon}, \ \mathbf{J}_f = \begin{bmatrix} \mathbf{J}_{f_1} \\ \mathbf{J}_{f_2}^* \end{bmatrix}, \ \mathbf{\Upsilon} = \begin{bmatrix} \mathbf{\Upsilon}_1 \\ \mathbf{\Upsilon}_2 \end{bmatrix}. \tag{3.72}$$

In (3.72), $\mathbf{J}_{f_2}^* \in \mathbb{R}^{(n-m) \times n}$ is assumed to have full row rank and it should be determined as a function of the first task's function Jacobian matrix in such a way that the dynamics of the first and second task are decoupled in terms of control so that the control vector associated with the first task is not directly influencing the function vector associated with the second task, and opposite. Additionally, it is assumed that the augmented function Jacobian matrix $\mathbf{J}_f$ is a full rank matrix, i.e., $\det(\mathbf{J}_f) \neq 0$. This implies that dynamics (3.72) is described by a set of linearly independent equations.

Taking into account (3.6), the last equation can be written in the following form

$$\begin{bmatrix} \dot{\mathbf{f}}_1 \\ \dot{\mathbf{f}}_2 \end{bmatrix} = \begin{bmatrix} \mathbf{J}_{f_1} \\ \mathbf{J}^*_{f_2} \end{bmatrix} \left[\mathbf{u}_q - \mathbf{A}^{-1} \left(\mathbf{b} + \mathbf{g} + \mathbf{T}_{ext} \right) \right] + \boldsymbol{\Upsilon}. \tag{3.73}$$

The control acceleration in the configuration space $\mathbf{u}_q$ is related to the control vectors in the function space as

$$\mathbf{u}_q = \begin{bmatrix} \boldsymbol{\Omega}_1 & \boldsymbol{\Omega}_2 \end{bmatrix} \begin{bmatrix} \mathbf{u}_{f_1} \\ \mathbf{u}_{f_2} \end{bmatrix} \tag{3.74}$$

where $\mathbf{u}_{f_1} \in \mathbb{R}^{m \times 1}$ is the control vector associated with the higher priority task, while $\mathbf{u}_{f_2} \in \mathbb{R}^{(n-m) \times 1}$ is the control vector associated with the lower priority task. Matrices $\boldsymbol{\Omega}_1 \in \mathbb{R}^{n \times m}$ and $\boldsymbol{\Omega}_2 \in \mathbb{R}^{n \times (n-m)}$ will be determined in order to achieve the desired decoupling.

Dynamics (3.73) can now be written as

$$\begin{bmatrix} \dot{\mathbf{f}}_1 \\ \dot{\mathbf{f}}_2 \end{bmatrix} = \begin{bmatrix} \mathbf{J}_{f_1} \\ \mathbf{J}^*_{f_2} \end{bmatrix} \begin{bmatrix} \boldsymbol{\Omega}_1 & \boldsymbol{\Omega}_2 \end{bmatrix} \begin{bmatrix} \mathbf{u}_{f_1} \\ \mathbf{u}_{f_2} \end{bmatrix} - \mathbf{J}_f \mathbf{A}^{-1} \left(\mathbf{b} + \mathbf{g} + \mathbf{T}_{ext} \right) + \boldsymbol{\Upsilon}. \tag{3.75}$$

In a more compact form, the last equation becomes

$$\begin{bmatrix} \dot{\mathbf{f}}_1 \\ \dot{\mathbf{f}}_2 \end{bmatrix} = \begin{bmatrix} \mathbf{J}_{f_1} \boldsymbol{\Omega}_1 & \mathbf{J}_{f_1} \boldsymbol{\Omega}_2 \\ \mathbf{J}^*_{f_2} \boldsymbol{\Omega}_1 & \mathbf{J}^*_{f_2} \boldsymbol{\Omega}_2 \end{bmatrix} \begin{bmatrix} \mathbf{u}_{f_1} \\ \mathbf{u}_{f_2} \end{bmatrix} - \mathbf{J}_f \mathbf{A}^{-1} \left(\mathbf{b} + \mathbf{g} + \mathbf{T}_{ext} \right) + \boldsymbol{\Upsilon}. \tag{3.76}$$

In order to achieve the desired decoupling and direct control of the two function vectors by the control vectors $\mathbf{u}_{f_1}$ and $\mathbf{u}_{f_2}$, the following conditions should be satisfied:

a) $\mathbf{J}_{f_1} \boldsymbol{\Omega}_1 = \mathbf{I}^{m \times m}$,
b) $\mathbf{J}_{f_1} \boldsymbol{\Omega}_2 = \mathbf{0}^{m \times (n-m)}$,
c) $\mathbf{J}^*_{f_2} \boldsymbol{\Omega}_1 = \mathbf{0}^{(n-m) \times m}$,
d) $\mathbf{J}^*_{f_2} \boldsymbol{\Omega}_2 = \mathbf{I}^{(n-m) \times (n-m)}$.

It is necessary to find $\mathbf{J}^*_{f_2}$, $\boldsymbol{\Omega}_1$, and $\boldsymbol{\Omega}_2$ such that listed conditions hold. These conditions will be discussed one by one.

The condition a) requires that $\boldsymbol{\Omega}_1$ is a right pseudoinverse of the matrix $\mathbf{J}_{f_1}$. Therefore, it can be written

$$\boldsymbol{\Omega}_1 = \mathbf{J}^\#_{f_1}. \tag{3.77}$$

For pseudoinverse matrix calculation, a weighted pseudoinverse will be taken, so it will be

$$\boldsymbol{\Omega}_1 = \mathbf{J}^\#_{f_1} = \mathbf{A}^{-1} \mathbf{J}^{\mathrm{T}}_{f_1} \left(\mathbf{J}_{f_1} \mathbf{A}^{-1} \mathbf{J}^{\mathrm{T}}_{f_1} \right)^{-1}. \tag{3.78}$$

Let us know discuss the condition c). It will be assumed that matrix $\mathbf{J}^*_{f_2}$ is composed as

$$\mathbf{J}^*_{f_2} = \mathbf{J}_{f_2} \mathbf{N}_{f_1}, \quad \mathbf{N}_{f_1} = \mathbf{I}^{n \times n} - \mathbf{J}^\#_{f_1} \mathbf{J}_{f_1} \tag{3.79}$$

where $\mathbf{N}_{f_1}$ is the null space projection matrix of the matrix $\mathbf{J}_{f_1}$. Considering (3.78) and (3.79), the product $\mathbf{J}^*_{f_2} \boldsymbol{\Omega}_1$ is

$$\mathbf{J}^*_{f_2} \boldsymbol{\Omega}_1 = \mathbf{J}_{f_2} \left(\mathbf{I} - \mathbf{J}^\#_{f_1} \mathbf{J}_{f_1} \right) \mathbf{J}^\#_{f_1} = \mathbf{0} \tag{3.80}$$

since the right pseudoinverse matrix satisfies condition

$$\mathbf{J}_{f_1}^{\#} \mathbf{J}_{f_1} \mathbf{J}_{f_1}^{\#} = \mathbf{J}_{f_1}^{\#}. \tag{3.81}$$

Thus, the condition c) is satisfied for $\mathbf{J}_{f_2}^{*}$ given by (3.79). With $\mathbf{J}_{f_2}^{*}$ in this form, the dynamics of the lower priority task is described in the null space of the higher priority task.

The condition d) implies that $\boldsymbol{\Omega}_2$ has to be selected as a right pseudoinverse of the matrix $\mathbf{J}_{f_2}^{*}$. Therefore, it will be calculated as a weighted pseudoinverse as

$$\boldsymbol{\Omega}_2 = \mathbf{J}_{f_2}^{*\#} = \mathbf{A}^{-1} \mathbf{J}_{f_2}^{*\mathrm{T}} \left(\mathbf{J}_{f_2}^{*} \mathbf{A}^{-1} \mathbf{J}_{f_2}^{*\mathrm{T}} \right)^{-1}. \tag{3.82}$$

Considering (3.79), it can be as well written

$$\boldsymbol{\Omega}_2 = \mathbf{A}^{-1} \left(\mathbf{J}_{f_2} \mathbf{N}_{f_1} \right)^{\mathrm{T}} \left[\left(\mathbf{J}_{f_2} \mathbf{N}_{f_1} \right) \mathbf{A}^{-1} \left(\mathbf{J}_{f_2} \mathbf{N}_{f_1} \right)^{\mathrm{T}} \right]^{-1}. \tag{3.83}$$

When $\mathbf{J}_{f_2}^{*}$, $\boldsymbol{\Omega}_1$ and $\boldsymbol{\Omega}_2$ are all defined, it is now necessary to check the condition b) given by $\mathbf{J}_{f_1} \boldsymbol{\Omega}_2 = \mathbf{0}$. Considering (3.83), this product is

$$\mathbf{J}_{f_1} \boldsymbol{\Omega}_2 = \mathbf{J}_{f_1} \mathbf{A}^{-1} \mathbf{N}_{f_1}^{\mathrm{T}} \mathbf{J}_{f_2}^{\mathrm{T}} \left[\left(\mathbf{J}_{f_2} \mathbf{N}_{f_1} \right) \mathbf{A}^{-1} \left(\mathbf{J}_{f_2} \mathbf{N}_{f_1} \right)^{\mathrm{T}} \right]^{-1}. \tag{3.84}$$

The term $\mathbf{J}_{f_1} \mathbf{A}^{-1} \mathbf{N}_{f_1}^{\mathrm{T}}$ will be analyzed. Since $\mathbf{N}_{f_1}$ is given in (3.79) this term becomes

$$\mathbf{J}_{f_1} \mathbf{A}^{-1} \mathbf{N}_{f_1}^{\mathrm{T}} = \mathbf{J}_{f_1} \mathbf{A}^{-1} \left(\mathbf{I} - \mathbf{J}_{f_1}^{\mathrm{T}} \mathbf{J}_{f_1}^{\#\mathrm{T}} \right). \tag{3.85}$$

Taking into account $\mathbf{J}_{f_1}^{\#}$ calculated in (3.78), it can be further shown

$$\mathbf{J}_{f_1} \mathbf{A}^{-1} \mathbf{N}_{f_1}^{\mathrm{T}} = \mathbf{J}_{f_1} \mathbf{A}^{-1} - \underbrace{\mathbf{J}_{f_1} \mathbf{A}^{-1} \mathbf{J}_{f_1}^{\mathrm{T}} \left(\mathbf{J}_{f_1} \mathbf{A}^{-1} \mathbf{J}_{f_1}^{\mathrm{T}} \right)^{-1}}_{\mathbf{I}} \mathbf{J}_{f_1} \mathbf{A}^{-1} = \mathbf{0}. \tag{3.86}$$

Thus, the condition b) is also satisfied.

The final expressions for matrices $\mathbf{J}_{f_2}^{*}$, $\boldsymbol{\Omega}_1$ and $\boldsymbol{\Omega}_2$ can be given as

$$\boldsymbol{\Omega}_1 = \mathbf{J}_{f_1}^{\#}, \quad \mathbf{J}_{f_1}^{\#} = \mathbf{A}^{-1} \mathbf{J}_{f_1}^{\mathrm{T}} \left(\mathbf{J}_{f_1} \mathbf{A}^{-1} \mathbf{J}_{f_1}^{\mathrm{T}} \right)^{-1} \tag{3.87}$$

$$\mathbf{J}_{f_2}^{*} = \mathbf{J}_{f_2} \mathbf{N}_{f_1}, \quad \mathbf{N}_{f_1} = \mathbf{I} - \mathbf{J}_{f_1}^{\#} \mathbf{J}_{f_1} \tag{3.88}$$

$$\boldsymbol{\Omega}_2 = \mathbf{J}_{f_2}^{*\#}, \quad \mathbf{J}_{f_2}^{*\#} = \mathbf{A}^{-1} \mathbf{J}_{f_2}^{*\mathrm{T}} \left(\mathbf{J}_{f_2}^{*} \mathbf{A}^{-1} \mathbf{J}_{f_2}^{*\mathrm{T}} \right)^{-1}. \tag{3.89}$$

With these matrices, the decoupling in terms of control will be achieved and specified hierarchy will be enforced. In that case, dynamics (3.76) can be written in the following form

$$\begin{bmatrix} \dot{\mathbf{f}}_1 \\ \dot{\mathbf{f}}_2 \end{bmatrix} = \begin{bmatrix} \mathbf{u}_{f_1} \\ \mathbf{u}_{f_2} \end{bmatrix} - \mathbf{J}_f \mathbf{A}^{-1} \left(\mathbf{b} + \mathbf{g} + \mathbf{T}_{ext} \right) + \boldsymbol{\Upsilon}, \quad \mathbf{J}_f = \begin{bmatrix} \mathbf{J}_{f_1} \\ \mathbf{J}_{f_2} \left(\mathbf{I} - \mathbf{J}_{f_1}^{\#} \mathbf{J}_{f_1} \right) \end{bmatrix}. \tag{3.90}$$

The last equation represents the dynamics of the system in the function space, that is now defined by two tasks in a hierarchical structure.

3.4　Relationship between Constraints and Tasks

Previously, the dynamics of the system with a single task or hierarchical structure of two tasks was discussed. For the sake of completeness of this work, it is also necessary to analyze a case when certain constraints need to be satisfied, together with different tasks which should be executed. Assume that system (3.6) needs to be controlled so that static equation $\boldsymbol{\Xi}(\mathbf{q}) = \mathbf{0}$ is satisfied, where $\boldsymbol{\Xi}(\mathbf{q}) \in \mathbb{R}^{m_c \times 1}$ ($m_c < n$) represents a vector-valued function of the system configuration with finite first- and second-order time derivatives.

One can formulate the stated problem as a requirement to select the control input such that motion of the system (3.6) is constrained to the manifold

$$S = \left\{ \mathbf{q}, \dot{\mathbf{q}} : \boldsymbol{\Xi}(\mathbf{q}) = \mathbf{0}, \boldsymbol{\Xi}(\mathbf{q}) \in \mathbb{R}^{m_c \times 1} \right\}. \tag{3.91}$$

The manifold is given in the configuration space. When the motion is constrained to the manifold, m_c components of the configuration vector have a defined functional dependence which is governed by $\boldsymbol{\Xi}(\mathbf{q}) = \mathbf{0}$; thus, the resulting motion in the manifold can be described by the remaining components of the configuration vector. This allows decomposition of the system motion into motion in the constrained direction and motion in the unconstrained direction. Additionally, this shows that certain degrees of freedom exist after satisfying the constraints and they can be utilized to execute tasks in the system. As the dimension of the control vector in the configuration space is not equal to the dimension of the constraint manifold, there is redundancy in the control vector with respect to the control enforcement of the constraints.

In order to have a unified approach and discuss the first-order dynamics for the constraints, as it was for the functions, we will define the constraint vector $\boldsymbol{\Phi}$ as

$$\boldsymbol{\Phi} = \dot{\boldsymbol{\Xi}} = \frac{\partial \boldsymbol{\Xi}}{\partial \mathbf{q}} \dot{\mathbf{q}} = \mathbf{J}_\Phi \dot{\mathbf{q}} \tag{3.92}$$

where $\partial \boldsymbol{\Xi}/\partial \mathbf{q} = \mathbf{J}_\Phi \in \mathbb{R}^{m_c \times n}$ is the constraint Jacobian matrix which is assumed to have full row rank. The first-order dynamics of the vector $\boldsymbol{\Phi}$ is now given by

$$\dot{\boldsymbol{\Phi}} = \mathbf{J}_\Phi \ddot{\mathbf{q}} + \dot{\mathbf{J}}_\Phi \dot{\mathbf{q}} = \mathbf{J}_\Phi \dot{\mathbf{v}} + \dot{\mathbf{J}}_\Phi \dot{\mathbf{q}}. \tag{3.93}$$

It will be analyzed how introduced constraints can be maintained while selected tasks are being fulfilled by the system (3.6). In this analysis, the following assumptions will be made

- Constraints are defined by the manifold $\boldsymbol{\Xi}(\mathbf{q}) = \mathbf{0} \in \mathbb{R}^{m_c \times 1}$, where $\boldsymbol{\Xi}(\mathbf{q})$ is two times differentiable with respect to time, and constraint Jacobian matrix $\partial \boldsymbol{\Xi}/\partial \mathbf{q} = \mathbf{J}_\Phi \in \mathbb{R}^{m_c \times n}$.
- One of the two tasks is defined by the function vector $\mathbf{f}_1 \in \mathbb{R}^{m_1 \times 1}$, which is differentiable with respect to time, and function Jacobian matrix $\mathbf{J}_{f_1} \in \mathbb{R}^{m_1 \times n}$.

- The second task is defined by the function vector $\mathbf{f}_2 \in \mathbb{R}^{m_2 \times 1}$, that is differentiable with respect to time, and function Jacobian matrix $\mathbf{J}_{f_2} \in \mathbb{R}^{m_2 \times n}$.
- The constraints have the highest priority, meaning that their priority is higher than the priorities of both tasks. On the other hand, the first task's priority is higher than the priority of the second task.
- All matrices $\mathbf{J}_\Phi \in \mathbb{R}^{m_c \times n}$, $\mathbf{J}_{f_1} \in \mathbb{R}^{m_1 \times n}$, and $\mathbf{J}_{f_2} \in \mathbb{R}^{m_2 \times n}$ are assumed to have full row rank; thus, the constraints and tasks are linearly independent.
- Without loss of generality, the number of available configuration space degrees of freedom is such that the constraints and tasks can be implemented concurrently without free degrees of freedom left; therefore, $m_c + m_1 + m_2 = n$.

With the listed requirements and application of the so far discussed approach, the first-order dynamics of the vectors associated with constraints and tasks is given by

$$
\begin{bmatrix} \dot{\boldsymbol{\Phi}} \\ \dot{\mathbf{f}}_1 \\ \dot{\mathbf{f}}_2 \end{bmatrix} = \begin{bmatrix} \mathbf{J}_\Phi \\ \mathbf{J}^*_{f_1} \\ \mathbf{J}^*_{f_2} \end{bmatrix} \dot{\mathbf{v}} + \begin{bmatrix} \dot{\mathbf{J}}_\Phi \dot{\mathbf{q}} \\ \boldsymbol{\Upsilon}_1 \\ \boldsymbol{\Upsilon}_2 \end{bmatrix} = \mathbf{J}_f \dot{\mathbf{v}} + \boldsymbol{\Upsilon}, \ \mathbf{J}_f = \begin{bmatrix} \mathbf{J}_\Phi \\ \mathbf{J}^*_{f_1} \\ \mathbf{J}^*_{f_2} \end{bmatrix}, \ \boldsymbol{\Upsilon} = \begin{bmatrix} \dot{\mathbf{J}}_\Phi \dot{\mathbf{q}} \\ \boldsymbol{\Upsilon}_1 \\ \boldsymbol{\Upsilon}_2 \end{bmatrix}. \quad (3.94)
$$

Matrices $\mathbf{J}^*_{f_1} \in \mathbb{R}^{m_1 \times n}$ and $\mathbf{J}^*_{f_2} \in \mathbb{R}^{m_2 \times n}$ are assumed to have full row rank. They should be determined as functions of constraint and function Jacobian matrices in such a way that constraints and tasks are decoupled in terms of control. It is assumed that constraint-function Jacobian matrix $\mathbf{J}_f \in \mathbb{R}^{n \times n}$ has full rank, i.e., $\det(\mathbf{J}_f) \neq 0$.

Considering (3.6), the last equation becomes

$$
\begin{bmatrix} \dot{\boldsymbol{\Phi}} \\ \dot{\mathbf{f}}_1 \\ \dot{\mathbf{f}}_2 \end{bmatrix} = \begin{bmatrix} \mathbf{J}_\Phi \\ \mathbf{J}^*_{f_1} \\ \mathbf{J}^*_{f_2} \end{bmatrix} \left[\mathbf{u}_q - \mathbf{A}^{-1} \left(\mathbf{b} + \mathbf{g} + \mathbf{T}_{ext} \right) \right] + \boldsymbol{\Upsilon}. \quad (3.95)
$$

The control acceleration in the configuration space $\mathbf{u}_q$ is related to the control vectors in the constraint-function space as

$$
\mathbf{u}_q = \begin{bmatrix} \boldsymbol{\Omega}_\Phi & \boldsymbol{\Omega}_1 & \boldsymbol{\Omega}_2 \end{bmatrix} \begin{bmatrix} \mathbf{u}_\Phi \\ \mathbf{u}_{f_1} \\ \mathbf{u}_{f_2} \end{bmatrix}. \quad (3.96)
$$

In (3.96), $\mathbf{u}_\Phi \in \mathbb{R}^{m_c \times 1}$ is the control vector associated with the constraints, $\mathbf{u}_{f_1} \in \mathbb{R}^{m_1 \times 1}$ is the control vector associated with the higher priority task, while $\mathbf{u}_{f_2} \in \mathbb{R}^{m_2 \times 1}$ is the control vector associated with the lower priority task. Matrices $\boldsymbol{\Omega}_\Phi \in \mathbb{R}^{n \times m_c}$, $\boldsymbol{\Omega}_1 \in \mathbb{R}^{n \times m_1}$ and $\boldsymbol{\Omega}_2 \in \mathbb{R}^{n \times m_2}$ will be determined so that the decoupling in terms of control is enforced, in the same way as it was in the previous section.

Dynamics (3.95) can now be written in the following form

$$\begin{bmatrix} \dot{\Phi} \\ \dot{f}_1 \\ \dot{f}_2 \end{bmatrix} = \begin{bmatrix} \mathbf{J}_\Phi \\ \mathbf{J}_{f_1}^* \\ \mathbf{J}_{f_2}^* \end{bmatrix} \begin{bmatrix} \mathbf{\Omega}_\Phi & \mathbf{\Omega}_1 & \mathbf{\Omega}_2 \end{bmatrix} \begin{bmatrix} \mathbf{u}_\Phi \\ \mathbf{u}_{f_1} \\ \mathbf{u}_{f_2} \end{bmatrix} - \mathbf{J}_f \mathbf{A}^{-1} \left(\mathbf{b} + \mathbf{g} + \mathbf{T}_{ext} \right) + \mathbf{\Upsilon}. \quad (3.97)$$

In a more compact form, the last equation can be written as

$$\begin{bmatrix} \dot{\Phi} \\ \dot{f}_1 \\ \dot{f}_2 \end{bmatrix} = \begin{bmatrix} \mathbf{J}_\Phi \mathbf{\Omega}_\Phi & \mathbf{J}_\Phi \mathbf{\Omega}_1 & \mathbf{J}_\Phi \mathbf{\Omega}_2 \\ \mathbf{J}_{f_1}^* \mathbf{\Omega}_\Phi & \mathbf{J}_{f_1}^* \mathbf{\Omega}_1 & \mathbf{J}_{f_1}^* \mathbf{\Omega}_2 \\ \mathbf{J}_{f_2}^* \mathbf{\Omega}_\Phi & \mathbf{J}_{f_2}^* \mathbf{\Omega}_1 & \mathbf{J}_{f_2}^* \mathbf{\Omega}_2 \end{bmatrix} \begin{bmatrix} \mathbf{u}_\Phi \\ \mathbf{u}_{f_1} \\ \mathbf{u}_{f_2} \end{bmatrix} - \mathbf{J}_f \mathbf{A}^{-1} \left(\mathbf{b} + \mathbf{g} + \mathbf{T}_{ext} \right) + \mathbf{\Upsilon}.$$
$$(3.98)$$

In order to achieve the desired decoupling and control of the constraint vector and two function vectors by the control vectors $\mathbf{u}_\Phi$, $\mathbf{u}_{f_1}$, and $\mathbf{u}_{f_2}$, the following conditions have to be satisfied:

a) $\mathbf{J}_\Phi \mathbf{\Omega}_\Phi = \mathbf{I}^{m_c \times m_c}$,
b) $\mathbf{J}_\Phi \mathbf{\Omega}_1 = \mathbf{0}^{m_c \times m_1}$,
c) $\mathbf{J}_\Phi \mathbf{\Omega}_2 = \mathbf{0}^{m_c \times m_2}$,
d) $\mathbf{J}_{f_1}^* \mathbf{\Omega}_\Phi = \mathbf{0}^{m_1 \times m_c}$,
e) $\mathbf{J}_{f_1}^* \mathbf{\Omega}_1 = \mathbf{I}^{m_1 \times m_1}$,
f) $\mathbf{J}_{f_1}^* \mathbf{\Omega}_2 = \mathbf{0}^{m_1 \times m_2}$,
g) $\mathbf{J}_{f_2}^* \mathbf{\Omega}_\Phi = \mathbf{0}^{m_2 \times m_c}$,
h) $\mathbf{J}_{f_2}^* \mathbf{\Omega}_1 = \mathbf{0}^{m_2 \times m_1}$,
i) $\mathbf{J}_{f_2}^* \mathbf{\Omega}_2 = \mathbf{I}^{m_2 \times m_2}$.

It is now necessary to find $\mathbf{J}_{f_1}^*$, $\mathbf{J}_{f_2}^*$, $\mathbf{\Omega}_\Phi$, $\mathbf{\Omega}_1$, and $\mathbf{\Omega}_2$ such that listed conditions hold.

For condition a) to be satisfied, matrix $\mathbf{\Omega}_\Phi$ needs to be selected as a right pseudoinverse of the matrix $\mathbf{J}_\Phi$. Here, a weighted pseudoinverse will be selected, so it can be written

$$\mathbf{\Omega}_\Phi = \mathbf{J}_\Phi^\# = \mathbf{A}^{-1} \mathbf{J}_\Phi^{\mathrm{T}} \left(\mathbf{J}_\Phi \mathbf{A}^{-1} \mathbf{J}_\Phi^{\mathrm{T}} \right)^{-1}. \quad (3.99)$$

Condition d) will now be discussed. Let us assume that matrix $\mathbf{J}_{f_1}^*$ is composed as

$$\mathbf{J}_{f_1}^* = \mathbf{J}_{f_1} \mathbf{N}_\Phi, \quad \mathbf{N}_\Phi = \mathbf{I}^{n \times n} - \mathbf{J}_\Phi^\# \mathbf{J}_\Phi \quad (3.100)$$

where $\mathbf{N}_\Phi$ is the null space projection matrix of the matrix $\mathbf{J}_\Phi$. Considering (3.99) and (3.100), the product $\mathbf{J}_{f_1}^* \mathbf{\Omega}_\Phi$ is

$$\mathbf{J}_{f_1}^* \mathbf{\Omega}_\Phi = \mathbf{J}_{f_1} \left(\mathbf{I} - \mathbf{J}_\Phi^\# \mathbf{J}_\Phi \right) \mathbf{J}_\Phi^\# = \mathbf{J}_{f_1} \left(\mathbf{J}_\Phi^\# - \mathbf{J}_\Phi^\# \underbrace{\mathbf{J}_\Phi \mathbf{J}_\Phi^\#}_{\mathbf{I}} \right) = \mathbf{J}_{f_1} \left(\mathbf{J}_\Phi^\# - \mathbf{J}_\Phi^\# \right) = 0$$
$$(3.101)$$

which shows that condition d) is satisfied for $\mathbf{J}_{f_1}^*$ selected as in (3.100). With $\mathbf{J}_{f_1}^*$ in this form, the dynamics of the first task is described in the null space of the constraints.

Condition e) implies that $\boldsymbol{\Omega}_1$ is a right pseudoinverse of the matrix $\mathbf{J}^*_{f_1}$. Thus, it will be calculated as a weighted pseudoinverse as follows

$$\boldsymbol{\Omega}_1 = \mathbf{J}^{*\#}_{f_1} = \mathbf{A}^{-1} \mathbf{J}^{*\mathrm{T}}_{f_1} \left(\mathbf{J}^*_{f_1} \mathbf{A}^{-1} \mathbf{J}^{*\mathrm{T}}_{f_1} \right)^{-1}. \tag{3.102}$$

Considering (3.100), (3.102) can also be written in the following form

$$\boldsymbol{\Omega}_1 = \mathbf{J}^{*\#}_{f_1} = \mathbf{A}^{-1} \left(\mathbf{J}_{f_1} \mathbf{N}_\Phi \right)^{\mathrm{T}} \left[\left(\mathbf{J}_{f_1} \mathbf{N}_\Phi \right) \mathbf{A}^{-1} \left(\mathbf{J}_{f_1} \mathbf{N}_\Phi \right)^{\mathrm{T}} \right]^{-1}. \tag{3.103}$$

Let us now check condition b). With $\boldsymbol{\Omega}_1$ given in (3.103), the product $\mathbf{J}_\Phi \boldsymbol{\Omega}_1$ becomes equal to

$$\mathbf{J}_\Phi \boldsymbol{\Omega}_1 = \mathbf{J}_\Phi \mathbf{A}^{-1} \mathbf{N}^{\mathrm{T}}_\Phi \mathbf{J}^{\mathrm{T}}_{f_1} \left[\left(\mathbf{J}_{f_1} \mathbf{N}_\Phi \right) \mathbf{A}^{-1} \left(\mathbf{J}_{f_1} \mathbf{N}_\Phi \right)^{\mathrm{T}} \right]^{-1}. \tag{3.104}$$

The term $\mathbf{J}_\Phi \mathbf{A}^{-1} \mathbf{N}^{\mathrm{T}}_\Phi$ will now be analyzed. With the $\mathbf{N}_\Phi$ matrix given by (3.100), this product is

$$\mathbf{J}_\Phi \mathbf{A}^{-1} \mathbf{N}^{\mathrm{T}}_\Phi = \mathbf{J}_\Phi \mathbf{A}^{-1} \left(\mathbf{I} - \mathbf{J}^{\mathrm{T}}_\Phi \mathbf{J}^{\#\mathrm{T}}_\Phi \right) \tag{3.105}$$

while the usage of $\mathbf{J}^{\#}_\Phi$ from (3.99) gives

$$\mathbf{J}_\Phi \mathbf{A}^{-1} \mathbf{N}^{\mathrm{T}}_\Phi = \mathbf{J}_\Phi \mathbf{A}^{-1} - \underbrace{\mathbf{J}_\Phi \mathbf{A}^{-1} \mathbf{J}^{\mathrm{T}}_\Phi \left(\mathbf{J}_\Phi \mathbf{A}^{-1} \mathbf{J}^{\mathrm{T}}_\Phi \right)^{-1}}_{\mathbf{I}} \mathbf{J}_\Phi \mathbf{A}^{-1} = 0. \tag{3.106}$$

Therefore, condition b) is satisfied for matrices $\mathbf{J}_\Phi$ and $\boldsymbol{\Omega}_1$ defined previously.

Conditions g) and h) lead to the selection of matrix $\mathbf{J}^*_{f_2}$ in the following form

$$\begin{aligned} \mathbf{J}^*_{f_2} \boldsymbol{\Omega}_\Phi &= \mathbf{0}^{m_2 \times m_c} \Rightarrow \mathbf{J}^*_{f_2} = \mathbf{J}_{f_2} \mathbf{N}_\Phi, \ \mathbf{N}_\Phi = \mathbf{I}^{n \times n} - \mathbf{J}^{\#}_\Phi \mathbf{J}_\Phi \\ \mathbf{J}^*_{f_2} \boldsymbol{\Omega}_1 &= \mathbf{0}^{m_2 \times m_1} \Rightarrow \mathbf{J}^*_{f_2} = \mathbf{J}_{f_2} \mathbf{N}_{f_1}, \ \mathbf{N}_{f_1} = \mathbf{I}^{n \times n} - \mathbf{J}^{*\#}_{f_1} \mathbf{J}^*_{f_1}. \end{aligned} \tag{3.107}$$

The matrices $\mathbf{N}_\Phi$ and $\mathbf{N}_{f_1}$ stand for the null space projection matrices of the constraint Jacobian and function Jacobian of the first task, respectively. The last equation can be rearranged as follows

$$\begin{aligned} \mathbf{J}^*_{f_2} \boldsymbol{\Omega}_\Phi &= \mathbf{J}_{f_2} \left(\mathbf{I}^{n \times n} - \mathbf{J}^{\#}_\Phi \mathbf{J}_\Phi \right) \boldsymbol{\Omega}_\Phi = \mathbf{0}^{m_2 \times m_c} \\ \mathbf{J}^*_{f_2} \boldsymbol{\Omega}_1 &= \mathbf{J}_{f_2} \left(\mathbf{I}^{n \times n} - \mathbf{J}^{*\#}_{f_1} \mathbf{J}^*_{f_1} \right) \boldsymbol{\Omega}_1 = \mathbf{0}^{m_2 \times m_1}. \end{aligned} \tag{3.108}$$

The null space projection matrices $\mathbf{N}_\Phi$ and $\mathbf{N}_{f_1}$ have the same dimensions but they are not identical. Accordingly, the structure of the matrix $\mathbf{J}^*_{f_2}$ cannot be derived directly from (3.108). Further transformation is needed to make the structure of the expressions in (3.108) in the form $\mathbf{J}_{f_2} \mathbf{N}_{\Phi f_1} \boldsymbol{\Omega}_\Phi$ and $\mathbf{J}_{f_2} \mathbf{N}_{\Phi f_1} \boldsymbol{\Omega}_1$, from where it could be deducted $\mathbf{J}^*_{f_2} = \mathbf{J}_{f_2} \mathbf{N}_{\Phi f_1}$. The transformation is undertaken by adding $-\mathbf{J}_{f_2} \mathbf{J}^{*\#}_{f_1} \mathbf{J}^*_{f_1} \boldsymbol{\Omega}_\Phi = \mathbf{0}^{m_2 \times m_c}$ to the first row in (3.108), considering condition d) from which $\mathbf{J}^*_{f_1}\mathbf{\Omega}_\Phi = \mathbf{0}^{m_1 \times m_c}$, and by adding $-\mathbf{J}_{f_2}\mathbf{J}^{\#}_\Phi\mathbf{J}_\Phi\mathbf{\Omega}_1 = \mathbf{0}^{m_2 \times m_1}$ to the second row in (3.108), taking into account condition b) implying $\mathbf{J}_\Phi\mathbf{\Omega}_1 = \mathbf{0}^{m_c \times m_1}$. After this transformation, one can write

$$
\begin{aligned}
\mathbf{J}^*_{f_2}\mathbf{\Omega}_\Phi &= \mathbf{J}_{f_2}\left(\mathbf{I}^{n \times n} - \mathbf{J}^{\#}_\Phi\mathbf{J}_\Phi - \mathbf{J}^{*\#}_{f_1}\mathbf{J}^*_{f_1}\right)\mathbf{\Omega}_\Phi = \mathbf{0}^{m_2 \times m_c} \\
\mathbf{J}^*_{f_2}\mathbf{\Omega}_1 &= \mathbf{J}_{f_2}\left(\mathbf{I}^{n \times n} - \mathbf{J}^{\#}_\Phi\mathbf{J}_\Phi - \mathbf{J}^{*\#}_{f_1}\mathbf{J}^*_{f_1}\right)\mathbf{\Omega}_1 = \mathbf{0}^{m_2 \times m_1}.
\end{aligned}
\tag{3.109}
$$

Thus, it follows $\mathbf{N}_{\Phi f_1} = \mathbf{I}^{n \times n} - \mathbf{J}^{\#}_\Phi\mathbf{J}_\Phi - \mathbf{J}^{*\#}_{f_1}\mathbf{J}^*_{f_1}$, and the matrix $\mathbf{J}^*_{f_2}$ is expressed as

$$
\mathbf{J}^*_{f_2} = \mathbf{J}_{f_2}\mathbf{N}_{\Phi f_1}, \quad \mathbf{N}_{\Phi f_1} = \mathbf{I}^{n \times n} - \mathbf{J}^{\#}_\Phi\mathbf{J}_\Phi - \mathbf{J}^{*\#}_{f_1}\mathbf{J}^*_{f_1}.
\tag{3.110}
$$

The matrix $\mathbf{J}^*_{f_2}$ given by (3.110) ensures that dynamics of the second task is described in the null space of the constraints and null space of the first task.

From condition i), it follows that $\mathbf{\Omega}_2$ is a right pseudoinverse of the matrix $\mathbf{J}^*_{f_2}$. Thus, it will be calculated as a weighted pseudoinverse by

$$
\mathbf{\Omega}_2 = \mathbf{J}^{*\#}_{f_2} = \mathbf{A}^{-1}\mathbf{J}^{*\mathrm{T}}_{f_2}\left(\mathbf{J}^*_{f_2}\mathbf{A}^{-1}\mathbf{J}^{*\mathrm{T}}_{f_2}\right)^{-1}.
\tag{3.111}
$$

Considering (3.110), (3.111) can be written in an alternative form as

$$
\mathbf{\Omega}_2 = \mathbf{J}^{*\#}_{f_2} = \mathbf{A}^{-1}\left(\mathbf{J}_{f_2}\mathbf{N}_{\Phi f_1}\right)^{\mathrm{T}}\left[\left(\mathbf{J}_{f_2}\mathbf{N}_{\phi f_1}\right)\mathbf{A}^{-1}\left(\mathbf{J}_{f_2}\mathbf{N}_{\phi f_1}\right)^{\mathrm{T}}\right]^{-1}.
\tag{3.112}
$$

It is necessary to check condition c). With $\mathbf{\Omega}_2$ from (3.112), the product $\mathbf{J}_\Phi\mathbf{\Omega}_2$ is

$$
\mathbf{J}_\Phi\mathbf{\Omega}_2 = \mathbf{J}_\Phi\mathbf{A}^{-1}\mathbf{N}^{\mathrm{T}}_{\Phi f_1}\mathbf{J}^{\mathrm{T}}_{f_2}\left[\left(\mathbf{J}_{f_2}\mathbf{N}_{\phi f_1}\right)\mathbf{A}^{-1}\left(\mathbf{J}_{f_2}\mathbf{N}_{\phi f_1}\right)^{\mathrm{T}}\right]^{-1}.
\tag{3.113}
$$

The term $\mathbf{J}_\Phi\mathbf{A}^{-1}\mathbf{N}^{\mathrm{T}}_{\Phi f_1}$ will be analyzed. Taking (3.110) into account, this term is

$$
\mathbf{J}_\Phi\mathbf{A}^{-1}\mathbf{N}^{\mathrm{T}}_{\Phi f_1} = \mathbf{J}_\Phi\mathbf{A}^{-1}\left(\mathbf{I} - \mathbf{J}^{\mathrm{T}}_\Phi\mathbf{J}^{\#\mathrm{T}}_\Phi - \mathbf{J}^{*\mathrm{T}}_{f_1}\mathbf{J}^{*\#\mathrm{T}}_{f_1}\right).
\tag{3.114}
$$

Considering (3.100), (3.114) can be further written as

$$
\mathbf{J}_\Phi\mathbf{A}^{-1}\mathbf{N}^{\mathrm{T}}_{\Phi f_1} = \mathbf{J}_\Phi\mathbf{A}^{-1}\mathbf{N}^{\mathrm{T}}_\Phi - \mathbf{J}_\Phi\mathbf{A}^{-1}\mathbf{N}^{\mathrm{T}}_\Phi\mathbf{J}^{\mathrm{T}}_{f_1}\mathbf{J}^{*\#\mathrm{T}}_{f_1}.
\tag{3.115}
$$

Finally, since it is proved $\mathbf{J}_\Phi\mathbf{A}^{-1}\mathbf{N}^{\mathrm{T}}_\Phi = \mathbf{0}$ in (3.106), then it follows

$$
\mathbf{J}_\Phi\mathbf{A}^{-1}\mathbf{N}^{\mathrm{T}}_{\Phi f_1} = \underbrace{\mathbf{J}_\Phi\mathbf{A}^{-1}\mathbf{N}^{\mathrm{T}}_\Phi}_{0} - \underbrace{\mathbf{J}_\Phi\mathbf{A}^{-1}\mathbf{N}^{\mathrm{T}}_\Phi}_{0}\mathbf{J}^{\mathrm{T}}_{f_1}\mathbf{J}^{*\#\mathrm{T}}_{f_1} = \mathbf{0}.
\tag{3.116}
$$

Thus, condition c) is also satisfied.

The last condition to be checked is condition f). With $\mathbf{\Omega}_2$ given in (3.112), the product $\mathbf{J}^*_{f_1}\mathbf{\Omega}_2$ can be calculated as

$$
\mathbf{J}^*_{f_1}\mathbf{\Omega}_2 = \mathbf{J}^*_{f_1}\mathbf{A}^{-1}\mathbf{N}^{\mathrm{T}}_{\Phi f_1}\mathbf{J}^{\mathrm{T}}_{f_2}\left[\left(\mathbf{J}_{f_2}\mathbf{N}_{\phi f_1}\right)\mathbf{A}^{-1}\left(\mathbf{J}_{f_2}\mathbf{N}_{\phi f_1}\right)^{\mathrm{T}}\right]^{-1}.
\tag{3.117}
$$

The term $\mathbf{J}_{f_1}^* \mathbf{A}^{-1} \mathbf{N}_{\Phi f_1}^{\mathrm{T}}$ will be analyzed. Taking (3.110) into account, this term is

$$\mathbf{J}_{f_1}^* \mathbf{A}^{-1} \mathbf{N}_{\Phi f_1}^{\mathrm{T}} = \mathbf{J}_{f_1}^* \mathbf{A}^{-1} \left(\mathbf{I} - \mathbf{J}_{\Phi}^{\mathrm{T}} \mathbf{J}_{\Phi}^{\#\mathrm{T}} - \mathbf{J}_{f_1}^{*\mathrm{T}} \mathbf{J}_{f_1}^{*\#\mathrm{T}} \right). \tag{3.118}$$

With $\mathbf{J}_{f_1}^{*\#}$ given in (3.102), (3.118) becomes

$$\mathbf{J}_{f_1}^* \mathbf{A}^{-1} \mathbf{N}_{\Phi f_1}^{\mathrm{T}} = \mathbf{J}_{f_1}^* \mathbf{A}^{-1} - \underbrace{\mathbf{J}_{f_1}^* \mathbf{A}^{-1} \mathbf{J}_{f_1}^{*\mathrm{T}} \left(\mathbf{J}_{f_1}^* \mathbf{A}^{-1} \mathbf{J}_{f_1}^{*\mathrm{T}} \right)^{-1}}_{\mathbf{I}} \mathbf{J}_{f_1}^* \mathbf{A}^{-1}$$
$$- \mathbf{J}_{f_1}^* \mathbf{A}^{-1} \mathbf{J}_{\Phi}^{\mathrm{T}} \mathbf{J}_{\Phi}^{\#\mathrm{T}} = - \mathbf{J}_{f_1}^* \mathbf{A}^{-1} \mathbf{J}_{\Phi}^{\mathrm{T}} \mathbf{J}_{\Phi}^{\#\mathrm{T}}. \tag{3.119}$$

When (3.100) is taken into account, (3.119) becomes

$$\mathbf{J}_{f_1}^* \mathbf{A}^{-1} \mathbf{N}_{\Phi f_1}^{\mathrm{T}} = - \mathbf{J}_{f_1} \underbrace{\mathbf{N}_{\Phi} \mathbf{A}^{-1} \mathbf{J}_{\Phi}^{\mathrm{T}}}_{\left(\mathbf{J}_{\Phi} \mathbf{A}^{-1} \mathbf{N}_{\Phi}^{\mathrm{T}} \right)^{\mathrm{T}}} \mathbf{J}_{\Phi}^{\#\mathrm{T}}. \tag{3.120}$$

From (3.106), it is $\mathbf{N}_{\Phi} \mathbf{A}^{-1} \mathbf{J}_{\Phi}^{\mathrm{T}} = \left(\mathbf{J}_{\Phi} \mathbf{A}^{-1} \mathbf{N}_{\Phi}^{\mathrm{T}} \right)^{\mathrm{T}} = \mathbf{0}^{\mathrm{T}} = \mathbf{0}$, so it follows

$$\mathbf{J}_{f_1}^* \mathbf{A}^{-1} \mathbf{N}_{\Phi f_1}^{\mathrm{T}} = \mathbf{0}. \tag{3.121}$$

Equation (3.121) proves that condition f) is satisfied.

Final expressions for $\mathbf{J}_{f_1}^*$, $\mathbf{J}_{f_2}^*$, $\mathbf{\Omega}_{\Phi}$, $\mathbf{\Omega}_1$, and $\mathbf{\Omega}_2$ are

$$\mathbf{\Omega}_{\Phi} = \mathbf{J}_{\Phi}^{\#}, \quad \mathbf{J}_{\Phi}^{\#} = \mathbf{A}^{-1} \mathbf{J}_{\Phi}^{\mathrm{T}} \left(\mathbf{J}_{\Phi} \mathbf{A}^{-1} \mathbf{J}_{\Phi}^{\mathrm{T}} \right)^{-1} \tag{3.122}$$

$$\mathbf{J}_{f_1}^* = \mathbf{J}_{f_1} \mathbf{N}_{\Phi}, \quad \mathbf{N}_{\Phi} = \mathbf{I} - \mathbf{J}_{\Phi}^{\#} \mathbf{J}_{\Phi} \tag{3.123}$$

$$\mathbf{\Omega}_1 = \mathbf{J}_{f_1}^{*\#}, \quad \mathbf{J}_{f_1}^{*\#} = \mathbf{A}^{-1} \mathbf{J}_{f_1}^{*\mathrm{T}} \left(\mathbf{J}_{f_1}^* \mathbf{A}^{-1} \mathbf{J}_{f_1}^{*\mathrm{T}} \right)^{-1} \tag{3.124}$$

$$\mathbf{J}_{f_2}^* = \mathbf{J}_{f_2} \mathbf{N}_{\Phi f_1}, \quad \mathbf{N}_{\Phi f_1} = \mathbf{I} - \mathbf{J}_{\Phi}^{\#} \mathbf{J}_{\Phi} - \mathbf{J}_{f_1}^{*\#} \mathbf{J}_{f_1}^* \tag{3.125}$$

$$\mathbf{\Omega}_2 = \mathbf{J}_{f_2}^{*\#}, \quad \mathbf{J}_{f_2}^{*\#} = \mathbf{A}^{-1} \mathbf{J}_{f_2}^{*\mathrm{T}} \left(\mathbf{J}_{f_2}^* \mathbf{A}^{-1} \mathbf{J}_{f_2}^{*\mathrm{T}} \right)^{-1}. \tag{3.126}$$

With matrices defined as in (3.122)–(3.126), the desired decoupling is achieved and specified hierarchy will be enforced. In that case, dynamics (3.98) becomes

$$\begin{bmatrix} \dot{\mathbf{\Phi}} \\ \dot{\mathbf{f}}_1 \\ \dot{\mathbf{f}}_2 \end{bmatrix} = \begin{bmatrix} \mathbf{u}_{\Phi} \\ \mathbf{u}_{f_1} \\ \mathbf{u}_{f_2} \end{bmatrix} - \mathbf{J}_f \mathbf{A}^{-1} \left(\mathbf{b} + \mathbf{g} + \mathbf{T}_{ext} \right) + \mathbf{\Upsilon},$$
$$\mathbf{J}_f = \begin{bmatrix} \mathbf{J}_{\Phi} \\ \mathbf{J}_{f_1} \left(\mathbf{I} - \mathbf{J}_{\Phi}^{\#} \mathbf{J}_{\Phi} \right) \\ \mathbf{J}_{f_2} \left(\mathbf{I} - \mathbf{J}_{\Phi}^{\#} \mathbf{J}_{\Phi} - \mathbf{J}_{f_1}^{*\#} \mathbf{J}_{f_1}^* \right) \end{bmatrix}, \quad \mathbf{\Upsilon} = \begin{bmatrix} \dot{\mathbf{J}}_{\Phi} \dot{\mathbf{q}} \\ \mathbf{\Upsilon}_1 \\ \mathbf{\Upsilon}_2 \end{bmatrix}. \tag{3.127}$$

The last equation is describing the dynamics of the system in the constraint-function space.

3.5 Control Synthesis

In this section, it will be shown how a control algorithm can be synthesized for the system dynamics derived in Sections 3.2, 3.3, and 3.4. It will be demonstrated how the algorithm can be designed using the configuration space control discussed in Section 3.1. After that, other approaches for synthesis directly in the function space will be presented. We will start by discussing control approaches for a single task, and then it will be shown how they can as well be utilized for more general cases. The main goal of the control algorithm is to enforce that $\mathbf{f}$ tracks its reference $\mathbf{f}^{ref}$ that is a differentiable vector-valued function of time, and will be denoted as the function vector reference. Therefore, the control goal can be expressed as

$$\mathbf{f} \xrightarrow{t \to \infty} \mathbf{f}^{ref}(t). \qquad (3.128)$$

Selection of the function vector and its reference will be discussed in detail in the next chapter.

3.5.1 Synthesis Based on Desired Configuration in the Configuration Space

Dynamics of the function vector for a non-redundant task and a redundant task can be described in the same form given by (3.32), but with the dimensions of the matrix $\mathbf{J}_f$, vector $\mathbf{f}$, and vector $\mathbf{\Upsilon}$ being different. For the non-redundant task, the matrix $\mathbf{J}_f$ is a square matrix, which is not the case for the non-redundant task. Thus, it will be assumed that $\mathbf{f}$ is an m-dimensional vector $m \leq n$.

In order to control the dynamics of the functions one can find the desired acceleration in the configuration space $\dot{\mathbf{v}}^{des} \in \mathbb{R}^{n \times 1}$. After that, the input force vector can be selected to enforce (3.7). This was discussed in Section 3.1. For calculation of $\dot{\mathbf{v}}^{des}$, one would have to select the desired dynamics of the function vector. The desired dynamics of the function vector will be denoted as $\dot{\mathbf{f}}^{des} \in \mathbb{R}^{m \times 1}$. In general, it can be expressed as

$$\dot{\mathbf{f}}^{des} = -\mathbf{\Psi}\left(\mathbf{f}, \mathbf{f}^{ref}, \dot{\mathbf{f}}^{ref}\right) \qquad (3.129)$$

where $\mathbf{\Psi}\left(\mathbf{f}, \mathbf{f}^{ref}, \dot{\mathbf{f}}^{ref}\right)$ is an arbitrary function selected such that

$$\dot{\mathbf{f}} + \mathbf{\Psi}\left(\mathbf{f}, \mathbf{f}^{ref}, \dot{\mathbf{f}}^{ref}\right) = 0 \qquad (3.130)$$

defines the desired dynamics of the vector $\mathbf{f}$. After selection of this function, it is necessary to calculate configuration space acceleration that will enforce $\dot{\mathbf{f}} = \dot{\mathbf{f}}^{des}$. This can be done if $\dot{\mathbf{f}}$ is replaced with $\dot{\mathbf{f}}^{des}$ in (3.32), which gives

$$\dot{\mathbf{f}}^{des} = \mathbf{J}_f \dot{\mathbf{v}} + \mathbf{\Upsilon}. \qquad (3.131)$$

If acceleration $\dot{\mathbf{v}}$ can be calculated from (3.131), it will represent desired acceleration in the configuration space. One can write (3.131) as follows

$$\mathbf{J}_f\dot{\mathbf{v}} = \dot{\mathbf{f}}^{des} - \boldsymbol{\Upsilon}. \tag{3.132}$$

where $\dot{\mathbf{v}}$ is the vector of n unknowns and $\mathbf{J}_f$ can be considered as the coefficient matrix of the linear system of equations (3.132). From (3.132), it is possible to calculate $\dot{\mathbf{v}}$ for any selected $\dot{\mathbf{f}}^{des}$ if and only if $\mathbf{J}_f$ is a full row rank matrix, which follows directly from the Kronecker–Capelli theorem. In the case when $m = n$, one can uniquely find desired configuration space acceleration. For $m < n$, a right pseudoinverse of $\mathbf{J}_f$ can be used for calculation of the desired configuration space acceleration. Therefore, the desired acceleration can be expressed as

$$\dot{\mathbf{v}}^{des} = \begin{cases} \mathbf{J}_f^{-1}\left(\dot{\mathbf{f}}^{des} - \boldsymbol{\Upsilon}\right), & m = n \\ \mathbf{J}_f^{\#}\left(\dot{\mathbf{f}}^{des} - \boldsymbol{\Upsilon}\right) + \boldsymbol{\Gamma}\ddot{\mathbf{q}}_0, & m < n \end{cases} \tag{3.133}$$

where $\mathbf{I} - \mathbf{J}_f^{\#}\mathbf{J}_f = \boldsymbol{\Gamma} \in \mathbb{R}^{n \times n}$ and $\ddot{\mathbf{q}}_0$ is an arbitrary acceleration vector in $\mathbb{R}^{n \times 1}$.

Once the desired acceleration in the configuration space is available, the approach from Section 3.1 is utilized to achieve (3.7).

3.5.2   Synthesis in the Function Space

First, control of the system with one non-redundant task will be considered, i.e., the design procedure will start from (3.41). The same approach can be applied for a redundant task, i.e., for dynamics (3.54). The only thing that will change are dimensions of vectors and matrices in the design process. The second and third term on the right hand side of (3.41) can be treated as the nonlinear disturbance vector denoted as $\mathbf{u}_{fdis} \in \mathbb{R}^{n \times 1}$. Therefore, (3.41) can be written as

$$\dot{\mathbf{f}} = \mathbf{u}_f - \underbrace{\left[\mathbf{J}_f\mathbf{A}^{-1}\left(\mathbf{b} + \mathbf{g} + \mathbf{T}_{ext}\right) - \boldsymbol{\Upsilon}\right]}_{\mathbf{u}_{fdis}} = \mathbf{u}_f - \mathbf{u}_{fdis}. \tag{3.134}$$

Clearly, (3.134) represents n first-order systems, for which many existing control methods can be applied.

3.5.2.1   Control Synthesis Based on Disturbance Estimation

One can notice a similarity between (3.9) and (3.134). Thus, the control synthesis can be done in a similar manner. Therefore, in this section, the control synthesis will be done in the control framework based on the disturbance observer. It is assumed that $\mathbf{f}(\mathbf{q}) \in \mathbb{R}^{n \times 1}$ needs to track the reference $\mathbf{f}^{ref}(t) \in \mathbb{R}^{n \times 1}$ which is a differentiable vector-valued function of time. Therefore, the control goal is expressed by (3.128).

If **f** is available (measured or calculated), the disturbance vector can be estimated using a classical disturbance observer for every component of the vector **f**, similarly as in (3.12). With the disturbance modeled as $\dot{\mathbf{u}}_{fdis} = \mathbf{0}$, the disturbance observer is constructed as follows

$$
\begin{aligned}
\dot{\mathbf{z}} &= \mathbf{L}\left(\mathbf{u}_f - \mathbf{z} + \mathbf{Lf}\right) \\
\hat{\mathbf{u}}_{fdis} &= \mathbf{z} - \mathbf{Lf}.
\end{aligned}
\tag{3.135}
$$

In (3.135), $\mathbf{L} \in \mathbb{R}^{n \times n}$ is again a constant gain matrix of the disturbance observer given by (3.10) and $\mathbf{u}_{fdis} + \mathbf{Lf} = \mathbf{z} \in \mathbb{R}^{n \times 1}$ is the intermediate variable in the disturbance estimation. The control goal will be achieved if $\mathbf{u}_f \in \mathbb{R}^{n \times 1}$ is calculated as

$$
\mathbf{u}_f = \hat{\mathbf{u}}_{fdis} + \dot{\mathbf{f}}^{ref} - \mathbf{K}_v \left(\mathbf{f} - \mathbf{f}^{ref}\right)
\tag{3.136}
$$

where $\mathbf{K}_v \in \mathbb{R}^{n \times n}$ is a constant diagonal matrix with positive diagonal entries given by (3.16). For $\mathbf{u}_f$ selected in this form, the closed-loop dynamics of the function vector is described by

$$
\left(\dot{\mathbf{f}} - \dot{\mathbf{f}}^{ref}\right) + \mathbf{K}_v \left(\mathbf{f} - \mathbf{f}^{ref}\right) = \hat{\mathbf{u}}_{fdis} - \mathbf{u}_{fdis}.
\tag{3.137}
$$

For the perfect disturbance estimation, i.e., $\hat{\mathbf{u}}_{fdis} = \mathbf{u}_{fdis}$, **f** exponentially converges to $\mathbf{f}^{ref}$ without overshoot.

The term $\dot{\mathbf{f}}^{ref} - \mathbf{K}_v \left(\mathbf{f} - \mathbf{f}^{ref}\right)$ can be identified as the desired dynamics of the function vector **f** and it can be denoted as $\dot{\mathbf{f}}^{des} \in \mathbb{R}^{n \times 1}$. Therefore, the control vector in the function space can be expressed as

$$
\mathbf{u}_f = \hat{\mathbf{u}}_{fdis} + \dot{\mathbf{f}}^{des}.
\tag{3.138}
$$

In general, one can select some other desired dynamics of the function vector, and the control vector in the function space will be calculated as in (3.138).

3.5.2.2 Control Synthesis Based on Sliding Mode Control

Control of the system (3.134) can be discussed using the sliding mode control theory [14, 28, 70, 71]. If the task is again to enforce the function vector to track the differentiable reference $\mathbf{f}^{ref}(t) \in \mathbb{R}^{n \times 1}$, the generalized error $\boldsymbol{\sigma} \in \mathbb{R}^{n \times 1}$ is defined as

$$
\boldsymbol{\sigma} = \begin{bmatrix} \sigma_1 & \sigma_2 & \cdots & \sigma_n \end{bmatrix}^{\mathrm{T}} = \mathbf{f} - \mathbf{f}^{ref}.
\tag{3.139}
$$

If the system motion is constrained to the manifold $\boldsymbol{\sigma} = \mathbf{0}$, the control goal will be achieved.

Considering (3.134) and (3.139), the first-order dynamics of the generalized error is

$$
\dot{\boldsymbol{\sigma}} = \mathbf{u}_f - \left(\mathbf{u}_{fdis} + \dot{\mathbf{f}}^{ref}\right).
\tag{3.140}
$$

If the equivalent control [70] $\mathbf{u}_f^{eq} \in \mathbb{R}^{n \times 1}$ is defined by

$$
\mathbf{u}_f^{eq} = \begin{bmatrix} u_{f1}^{eq} & u_{f2}^{eq} & \cdots & u_{fn}^{eq} \end{bmatrix}^{\mathrm{T}} = \mathbf{u}_{fdis} + \dot{\mathbf{f}}^{ref}
\tag{3.141}
$$

then (3.140) has the form

$$\dot{\boldsymbol{\sigma}} = \mathbf{u}_f - \mathbf{u}_f^{eq}. \tag{3.142}$$

A Lyapunov function candidate is selected as follows

$$V = \frac{\boldsymbol{\sigma}^{\mathrm{T}} \boldsymbol{\sigma}}{2}. \tag{3.143}$$

Using (3.142) and (3.143), the time derivative of the Lyapunov function candidate becomes

$$\dot{V} = \boldsymbol{\sigma}^{\mathrm{T}} \dot{\boldsymbol{\sigma}} = \boldsymbol{\sigma}^{\mathrm{T}} \left(\mathbf{u}_f - \mathbf{u}_f^{eq} \right). \tag{3.144}$$

Let us assume that equivalent control's components are bounded and they satisfy the following set of inequalities

$$\left| u_{fi}^{eq} \right| < M_i \left(\mathbf{q} \right), \ M_i \left(\mathbf{q} \right) > 0, \ i = 1, 2, \ldots, n. \tag{3.145}$$

By selecting the control vector in the function space as

$$\mathbf{u}_f = \begin{bmatrix} -M_1 \mathrm{sign} \left(\sigma_1 \right) & -M_2 \mathrm{sign} \left(\sigma_2 \right) & \cdots & -M_n \mathrm{sign} \left(\sigma_n \right) \end{bmatrix}^{\mathrm{T}} \tag{3.146}$$

it is possible to enforce $\dot{V} < 0$, meaning that V is becoming a negative definite Lyapunov function and the Lyapunov asymptotic stability criterion is satisfied. Thus, the control goal is achieved. For $\mathbf{u}_f$ from (3.146), the closed-loop dynamics of the generalized error is

$$\dot{\boldsymbol{\sigma}} + \begin{bmatrix} M_1 \mathrm{sign} \left(\sigma_1 \right) + u_{f1}^{eq} \\ M_2 \mathrm{sign} \left(\sigma_2 \right) + u_{f2}^{eq} \\ \vdots \\ M_n \mathrm{sign} \left(\sigma_n \right) + u_{fn}^{eq} \end{bmatrix} = \mathbf{0}, \ M_i \mathrm{sign} \left(\sigma_i \right) + u_{fi}^{eq} > 0, \ i = 1, 2, \ldots, n \tag{3.147}$$

and it ensures the finite time convergence to the manifold $\boldsymbol{\sigma} = \mathbf{0}$.

3.5.2.3   Control Synthesis Based on Equivalent Control Estimation

One more approach for control design, based on the equivalent control estimation [54], will be discussed. We will start from the first-order dynamics of the generalized error (3.142). If $\boldsymbol{\sigma}$ is available from measurement or calculation, equivalent control can be estimated from (3.142) applying the same approach applied for the disturbance estimation (3.12) and (3.135). The equivalent control is modeled by $\dot{\mathbf{u}}_f^{eq} = \mathbf{0}$, and the equivalent control observer is designed as

$$\begin{aligned} \dot{\mathbf{z}} &= \mathbf{L} \left(\mathbf{u}_f - \mathbf{z} + \mathbf{L} \boldsymbol{\sigma} \right) \\ \hat{\mathbf{u}}_f^{eq} &= \mathbf{z} - \mathbf{L} \boldsymbol{\sigma} \end{aligned} \tag{3.148}$$

where $\mathbf{L} \in \mathbb{R}^{n \times n}$ and $\mathbf{u}_f^{eq} + \mathbf{L} \boldsymbol{\sigma} = \mathbf{z} \in \mathbb{R}^{n \times 1}$ represent a constant gain matrix of the equivalent control observer given in (3.10) and the intermediate variable in the equivalent control estimation, respectively. The control vector in the function space can be selected as

$$\mathbf{u}_f = \hat{\mathbf{u}}_f^{eq} - \boldsymbol{\Psi}(\boldsymbol{\sigma}) \tag{3.149}$$

where $\boldsymbol{\Psi}(\boldsymbol{\sigma}) \in \mathbb{R}^{n \times 1}$ is a function selected in such a manner that

$$\dot{\boldsymbol{\sigma}} + \boldsymbol{\Psi}(\boldsymbol{\sigma}) = \mathbf{0} \tag{3.150}$$

defines the desired dynamics of the generalized error $\boldsymbol{\sigma}$. The desired dynamics needs to ensure $\boldsymbol{\sigma} \xrightarrow{t \to \infty} \mathbf{0}$. From (3.150), it follows that $-\boldsymbol{\Psi}(\boldsymbol{\sigma})$ represents the desired dynamics of the generalized error which can be denoted as $\dot{\boldsymbol{\sigma}}^{des}$. Therefore, the control vector in the function space can alternatively be written in the following form

$$\mathbf{u}_f = \hat{\mathbf{u}}_f^{eq} + \dot{\boldsymbol{\sigma}}^{des} \tag{3.151}$$

For $\mathbf{u}_f$ from (3.149), the resulting closed-loop dynamics of the generalized error is described by

$$\dot{\boldsymbol{\sigma}} + \boldsymbol{\Psi}(\boldsymbol{\sigma}) = \hat{\mathbf{u}}_f^{eq} - \mathbf{u}_f^{eq}. \tag{3.152}$$

For the perfect equivalent control estimation, i.e., $\hat{\mathbf{u}}_f^{eq} = \mathbf{u}_f^{eq}$, the resulting dynamics is equal to the desired dynamics of the generalized error described by (3.150).

The function $\boldsymbol{\Psi}(\boldsymbol{\sigma})$ can be selected such that exponential convergence to the manifold $\boldsymbol{\sigma} = \mathbf{0}$ is enforced, or as the one which will lead to the finite time convergence. For the exponential convergence, one can define $\boldsymbol{\Psi}(\boldsymbol{\sigma})$ as

$$\boldsymbol{\Psi}(\boldsymbol{\sigma}) = \mathbf{D}\boldsymbol{\sigma} \tag{3.153}$$

where $\mathbf{D} \in \mathbb{R}^{n \times n}$ is a constant diagonal matrix

$$\mathbf{D} = \text{diag}(d_1, d_2, \ldots, d_n), \ d_i > 0, \ i = 1, 2, \ldots, n. \tag{3.154}$$

If the finite time convergence is desired, then $\boldsymbol{\Psi}(\boldsymbol{\sigma})$ can be selected as

$$\boldsymbol{\Psi}(\boldsymbol{\sigma}) = \mathbf{D}\left[|\sigma_1|^{2\alpha-1}\text{sign}(\sigma_1) \quad \cdots \quad |\sigma_n|^{2\alpha-1}\text{sign}(\sigma_n)\right]^{\mathrm{T}}, 0.5 \le \alpha < 1. \tag{3.155}$$

For $\alpha = 0.5$, discontinuous control is selected and it will enforce the shortest finite convergence time to the manifold $\boldsymbol{\sigma} = \mathbf{0}$. If $0.5 < \alpha < 1$ is valid, then control will be continuous, but again with finite time convergence. Obviously, $\alpha = 1$ in (3.155) gives the same solution as in (3.153), i.e., exponential convergence is obtained.

3.5.2.4 Mapping of Control Vector from Function Space to Configuration Space

When control in the function space is calculated as (3.138), or (3.146), or (3.149), it is necessary to transform it back to the configuration space control acceleration $\mathbf{u}_q$, from which the input force vector $\mathbf{T}$ is easily obtainable. From (3.3), mapping (3.39) which holds for $m = n$, and mapping (3.50) which is valid for $m < n$, it follows that the transformation is done as

$$\mathbf{T} = \mathbf{A}\mathbf{u}_q, \ \mathbf{u}_q = \begin{cases} \mathbf{J}_f^{-1}\mathbf{u}_f, & m = n \\ \mathbf{J}_f^{\#}\mathbf{u}_f + \mathbf{\Gamma}\mathbf{u}_{q_0}, & m < n \end{cases} \tag{3.156}$$

where $\mathbf{I} - \mathbf{J}_f^{\#}\mathbf{J}_f = \mathbf{\Gamma} \in \mathbb{R}^{n \times n}$ and $\mathbf{u}_{q_0}$ is an arbitrary acceleration vector in $\mathbb{R}^{n \times 1}$. One can actually use the equation given for redundant task, $\mathbf{T} = \mathbf{A}\left(\mathbf{J}_f^{\#}\mathbf{u}_f + \mathbf{\Gamma}\mathbf{u}_{q_0}\right)$ as the general solution which is valid for $m \leq n$. For $m = n$ it generates the first equation in (3.156).

3.5.3   Control of System with Multiple Tasks

Control of the system with multiple tasks, presented in Section 3.3, will now be discussed. The dynamics of the two tasks is given by (3.72). If one wants to apply the approach given in Section 3.5.1, the desired dynamics for the two function vectors $\dot{\mathbf{f}}_1^{des}$ and $\dot{\mathbf{f}}_2^{des}$ are combined in the vector $\dot{\mathbf{f}}^{des} \in \mathbb{R}^{n \times 1}$ as

$$\dot{\mathbf{f}}^{des} = \begin{bmatrix} \dot{\mathbf{f}}_1^{des} \\ \dot{\mathbf{f}}_2^{des} \end{bmatrix}. \tag{3.157}$$

With this $\dot{\mathbf{f}}^{des}$, the same procedure explained in Section 3.5.1 can be applied.

If one wants to design the control algorithm in the function space, control of the system dynamics (3.90) needs to be discussed. The second and third term on the right hand side of (3.90) can be considered as the nonlinear disturbance vector. First m components of that vector can be denoted as $\mathbf{u}_{f_1 dis} \in \mathbb{R}^{m \times 1}$, while the remaining $(n - m)$ components will be represented as $\mathbf{u}_{f_2 dis} \in \mathbb{R}^{(n-m) \times 1}$. Then, (3.90) can be written in the form

$$\begin{bmatrix} \dot{\mathbf{f}}_1 \\ \dot{\mathbf{f}}_2 \end{bmatrix} = \begin{bmatrix} \mathbf{u}_{f_1} \\ \mathbf{u}_{f_2} \end{bmatrix} - \underbrace{\left[\mathbf{J}_f \mathbf{A}^{-1}\left(\mathbf{b} + \mathbf{g} + \mathbf{T}_{ext} \right) - \mathbf{\Upsilon} \right]}_{\left[\mathbf{u}_{f_1 dis} \ \mathbf{u}_{f_2 dis} \right]^{\mathrm{T}}} = \begin{bmatrix} \mathbf{u}_{f_1} \\ \mathbf{u}_{f_2} \end{bmatrix} - \begin{bmatrix} \mathbf{u}_{f_1 dis} \\ \mathbf{u}_{f_2 dis} \end{bmatrix}. \tag{3.158}$$

Consequently, (3.158) represents the dynamics of two systems written in the same form as (3.134). Therefore, the design procedure for $\mathbf{u}_{f_1}$ and $\mathbf{u}_{f_2}$ will be the one already presented for dynamics (3.134). When these control vectors in the function space are calculated, they need to be transformed back to the vector $\mathbf{T}$. Using (3.3) and (3.74), it is done as

$$\mathbf{T} = \mathbf{A}\left(\mathbf{\Omega}_1 \mathbf{u}_{f_1} + \mathbf{\Omega}_2 \mathbf{u}_{f_2} \right). \tag{3.159}$$

where $\mathbf{\Omega}_1$ and $\mathbf{\Omega}_2$ are given by (3.87)–(3.89).

3.5.4 Control of System with Constraints and Tasks

One needs to discuss control design for the system containing both constraints and tasks, the case presented in Section 3.4. The desired dynamics for the constraint vector $\dot{\boldsymbol{\Phi}}^{des}$ needs to be selected in such a way that convergence to the manifold (3.91) is enforced. In general it is given as

$$\dot{\boldsymbol{\Phi}}^{des} = -\boldsymbol{\Psi}\left(\boldsymbol{\Phi}, \boldsymbol{\Phi}^{ref}, \dot{\boldsymbol{\Phi}}^{ref}\right) \tag{3.160}$$

where $\boldsymbol{\Psi}\left(\boldsymbol{\Phi}, \boldsymbol{\Phi}^{ref}, \dot{\boldsymbol{\Phi}}^{ref}\right) \in \mathbb{R}^{m_c \times 1}$ is a function selected in such a manner that

$$\dot{\boldsymbol{\Phi}}^{des} + \boldsymbol{\Psi}\left(\boldsymbol{\Phi}, \boldsymbol{\Phi}^{ref}, \dot{\boldsymbol{\Phi}}^{ref}\right) = \mathbf{0} \tag{3.161}$$

defines a desired first-order dynamics of the constraint vector and convergence to its reference $\boldsymbol{\Phi}^{ref} \in \mathbb{R}^{m_c \times 1}$. The desired dynamics needs to ensure $\boldsymbol{\Phi} \xrightarrow{t \to \infty} \boldsymbol{\Phi}^{ref}$, which consequently guarantees convergence to the manifold (3.91).

A straightforward solution is to select

$$\dot{\boldsymbol{\Phi}}^{des} = \dot{\boldsymbol{\Phi}}^{ref} - \mathbf{K}_{\Phi v}\left(\boldsymbol{\Phi} - \boldsymbol{\Phi}^{ref}\right) \tag{3.162}$$

where $\mathbf{K}_{\Phi v} \in \mathbb{R}^{m_c \times m_c}$ is a constant diagonal matrix

$$\mathbf{K}_{\Phi v} = \text{diag}\left(k_{\Phi v 1}, k_{\Phi v 2}, \ldots, k_{\Phi v m_c}\right), \ k_{\Phi v i} > 0, \ i = 1, 2, \ldots, \ m_c. \tag{3.163}$$

The reference $\boldsymbol{\Phi}^{ref}$ can be chosen as follows

$$\boldsymbol{\Phi}^{ref} = -\mathbf{K}_{\Phi p}\boldsymbol{\Xi} \tag{3.164}$$

where $\mathbf{K}_{\Phi p} \in \mathbb{R}^{m_c \times m_c}$ is a constant diagonal matrix

$$\mathbf{K}_{\Phi p} = \text{diag}\left(k_{\Phi p 1}, k_{\Phi p 2}, \ldots, k_{\Phi p m_c}\right), \ k_{\Phi p i} > 0, \ i = 1, 2, \ldots, \ m_c. \tag{3.165}$$

This desired dynamics for the constraint vector ensures exponential convergence to the manifold (3.91).

If one wants to design control in the function space, it is necessary to discuss control of the system dynamics (3.127). The second and third term on the right hand side of the equation can be considered as the nonlinear disturbance vector. First m_c components of that vector can be denoted as $\mathbf{u}_{\Phi dis} \in \mathbb{R}^{m_c \times 1}$, $\mathbf{u}_{f_1 dis} \in \mathbb{R}^{m_1 \times 1}$ will stand for the next m_1 components, while $\mathbf{u}_{f_2 dis} \in \mathbb{R}^{m_2 \times 1}$ will represent the remaining m_2 components. With this notation, (3.127) becomes

$$\begin{bmatrix} \dot{\boldsymbol{\Phi}} \\ \dot{\mathbf{f}}_1 \\ \dot{\mathbf{f}}_2 \end{bmatrix} = \begin{bmatrix} \mathbf{u}_{\Phi} \\ \mathbf{u}_{f_1} \\ \mathbf{u}_{f_2} \end{bmatrix} - \underbrace{\left[\mathbf{J}_f \mathbf{A}^{-1}\left(\mathbf{b} + \mathbf{g} + \mathbf{T}_{ext}\right) - \boldsymbol{\Upsilon}\right]}_{\left[\mathbf{u}_{\Phi dis} \ \mathbf{u}_{f_1 dis} \ \mathbf{u}_{f_2 dis}\right]^{\mathrm{T}}} = \begin{bmatrix} \mathbf{u}_{\Phi} \\ \mathbf{u}_{f_1} \\ \mathbf{u}_{f_2} \end{bmatrix} - \begin{bmatrix} \mathbf{u}_{\Phi dis} \\ \mathbf{u}_{f_1 dis} \\ \mathbf{u}_{f_2 dis} \end{bmatrix}. \tag{3.166}$$

Thus, (3.166) represents the dynamics of three systems written in the same form as (3.134). Therefore, the design procedure for $\mathbf{u}_{\Phi dis}$, $\mathbf{u}_{f_1}$, and $\mathbf{u}_{f_2}$ will be the same as the one presented for dynamics (3.134). The reference for the constraint vector can be selected as (3.164). When the control vectors in the constraint-function space are calculated, they have to be transformed back to the input force vector $\mathbf{T}$. Considering (3.3) and (3.96), it is done by

$$\mathbf{T} = \mathbf{A} \left(\boldsymbol{\Omega}_\Phi \mathbf{u}_\Phi + \boldsymbol{\Omega}_1 \mathbf{u}_{f_1} + \boldsymbol{\Omega}_2 \mathbf{u}_{f_2} \right). \qquad (3.167)$$

where $\boldsymbol{\Omega}_\Phi$, $\boldsymbol{\Omega}_1$, and $\boldsymbol{\Omega}_2$ are given in (3.122)–(3.126).

3.5.5   Important Remarks about Control Synthesis

An important thing has to be noted here. In the control design, the inertia matrix $\mathbf{A}\,(\mathbf{q})$ is assumed to be known. If this is not the case, and only the nominal inertia matrix $\mathbf{A}_n\,(\mathbf{q})$ is known, then the inertia matrix can be written as $\mathbf{A}\,(\mathbf{q}) = \mathbf{A}_n\,(\mathbf{q}) + \Delta\mathbf{A}\,(\mathbf{q})$, where $\Delta\mathbf{A}\,(\mathbf{q})$ is the unknown variation of the inertia matrix. Then, dynamics (2.1) can be written as

$$\mathbf{A}_n \ddot{\mathbf{q}} + \mathbf{b} + \mathbf{g} + \mathbf{T}_{ext} + \Delta\mathbf{A}\ddot{\mathbf{q}} = \mathbf{T}. \qquad (3.168)$$

In (3.168), $\Delta\mathbf{A}\ddot{\mathbf{q}}$ can be treated in the same way as unknown forces $\mathbf{b}$, $\mathbf{g}$ and $\mathbf{T}_{ext}$. Therefore, the presented approach for the control design can still be applied. The only difference is that nominal inertia matrix $\mathbf{A}_n$ is used instead of the inertia matrix $\mathbf{A}$ in all places during the control design.

It is also very common to have a control system with layered structure. One can have a high-level controller which calculates references for low-level controllers which enforce tracking of these references in the controlled system. The high-level controller calculates the references based on controlled functions. Such an approach is illustrated for the formation control of mobile robots in Chapter 6. Another form of a control system with layered structure can exist if in the system (3.168) one compensates unknown forces in the configuration space, which can be considered as disturbance, and this can be treated as a low-level control action. Then, a control algorithm on a higher level can be designed for this compensated system. Such an approach will be used for a bilateral system in Section 8.2. In this system, only nominal inertias of the master and slave devices are assumed to be known. Therefore, this example will also address the issue discussed in the previous paragraph. Compensation of the configuration space disturbances and control of compensated systems are also adopted for an object manipulation task in three-dimensional space in Section 8.3.2.

There is one more remark that has to be given here. Throughout this chapter, when a right pseudoinverse had to be selected, the pseudoinverse matrix which minimizes the kinetic energy or 'acceleration energy' was chosen. However, it is not obligatory, and the approach presented in this chapter is also applicable with other right pseudoinverse matrices. For example, instead of the kinetic energy of the controlled system, one can select to minimize the function $K\,(\dot{\mathbf{q}}) = 0.5\dot{\mathbf{q}}^{\mathrm{T}}\mathbf{W}\dot{\mathbf{q}}$, where $\mathbf{W}$ is a symmetric positive definite matrix taken as a weighting matrix. In this case, the inertia matrix $\mathbf{A}$ is replaced with the matrix $\mathbf{W}$ which can be selected according to some criterion. In the approach presented in this chapter, one would only need to replace $\mathbf{A}$ with $\mathbf{W}$ in all calculations of a right pseudoinverse matrix, but the whole approach is still valid. Thus, selection of the right pseudoinverse brings an additional degree of freedom in the control design, once the number of the controlled functions is less than the dimension of the configuration vector. This will be demonstrated in an example in Chapter 6. In the systems where the force is treated as control input, the selection of the weighting matrix $\mathbf{W}$ is not free, since the matrix $\mathbf{A}$ has to be used as the weighting matrix if the dynamical decoupling is desired [54, 36].

3.6 Conclusion

In this chapter a novel approach to design of motion control systems has been discussed. The method is based on control of functions that describe a task that is to be executed by the controlled system. The approach relies on representing the dynamics of the system in a new space, called function space. In the configuration space, the control acceleration vector is considered as control signal, while the input force is taken just as a mean to enforce that acceleration through the control distribution matrix which is in this case the inverse inertia matrix. Using an appropriate mapping of the control signal in the function space back to the control acceleration vector in the configuration space, the control distribution matrix in the function space can be made to be an identity matrix. Thus, dynamics of each function can be enforced to have a desired form. Control synthesis can be done based on the desired acceleration in the configuration space, or it can be performed directly in the function space. In this chapter, three different control methods were proposed for the control synthesis in the function space: disturbance-observer-based control, sliding mode control, and control based on the equivalent control estimation. In addition, control of the system with hierarchical structure of multiple tasks, or the system in which certain constraints and tasks exist has been discussed. Using the same approach, control of the system with a redundant task has been also considered.

4

Functions in Motion Control Systems

Up to this point, it was assumed that functions in a motion control system are selected before the control algorithm synthesis begins. However, the process of selection of functions was not discussed. In this chapter, the selection will be considered in more detail. It is important to discuss selection and properties of the functions, so that control synthesis can be done by the approach presented in the previous chapter.

4.1 Selection of Function Vector and Selection of Function Vector Reference

In this discussion, it will be assumed that a mechanical system is described by (2.1) or (3.6). Further, we will assume that m functions ($m \leq n$) have to be controlled in the system. These functions are φ_i, $i = 1, 2, \ldots, m$, given in one of the forms (3.20)–(3.23). The functions φ_i can be combined in the vector $\boldsymbol{\varphi} = [\varphi_1 \ \varphi_2 \ldots \varphi_m]^{\mathrm{T}}$, while the vector $\boldsymbol{\varphi}^{ref}$ is formed from their references as $\boldsymbol{\varphi}^{ref} = \left[\varphi_1^{ref} \ \varphi_2^{ref} \ldots \varphi_m^{ref}\right]^{\mathrm{T}}$.

Based on the functions φ_i and their references φ_i^{ref}, one is selecting functions f_i, given in the vector form as the function vector $\mathbf{f}$, and their references f_i^{ref}, given in the vector form as the function vector reference $\mathbf{f}^{ref}$. A control law is then synthesized to enforce (3.128). As a consequence, one wants to impose

$$\boldsymbol{\varphi} \xrightarrow{t \to \infty} \boldsymbol{\varphi}^{ref}. \tag{4.1}$$

In Section 3.2.1, it was explained how functions f_i are formed based on the functions φ_i. Another question which has to be answered is how to select f_i^{ref} with the given φ_i^{ref}. For φ_i given by (3.20), we have assumed the form (3.24), which as well gives the relationship between f_i and φ_i. Thus, f_i^{ref} can be calculated as

$$f_i^{ref} = \int_0^t \varphi_i^{ref} \mathrm{d}t + f_i^{ref}(0) \tag{4.2}$$

where $f_i^{ref}(0)$ can be arbitrarily selected. If φ_i is given as in (3.21) or (3.22), f_i and φ_i are identical to each other, as indicated by (3.26). Naturally, their references are also identical and

$$f_i^{ref} = \varphi_i^{ref}. \tag{4.3}$$

When φ_i has the form (3.23), one needs to enforce that

$$f_i \xrightarrow{t \to \infty} f_i^{ref} \tag{4.4}$$

causes

$$\varphi_i \xrightarrow{t \to \infty} \varphi_i^{ref}. \tag{4.5}$$

A straightforward selection in that case is to have

$$f_i^{ref} = \dot{\varphi}_i^{ref} - c_i \left(\varphi_i - \varphi_i^{ref} \right), \quad c_i > 0. \tag{4.6}$$

An important issue to be considered is whether (3.128) can be satisfied at all. One can analyze the differential equation $\mathbf{f}(\mathbf{q}, \mathbf{v}) = \mathbf{f}^{ref}(t)$, and check whether it is possible to determine the configuration vector $\mathbf{q} \in \mathbb{R}^{n \times 1}$ from that equation. If it is not possible to find the configuration vector which satisfies the equation, then obviously (3.128) cannot be enforced and selection of functions or their references is not appropriate.

As an elementary example, one can consider the functions φ_1 and φ_2 along with their references given in the vector form as

$$\boldsymbol{\varphi}(\mathbf{q}) = \begin{bmatrix} q_1^2 + q_2^2 \\ q_1 + q_2 \end{bmatrix}, \quad \boldsymbol{\varphi}^{ref}(t) = \begin{bmatrix} 4 \\ 3 \end{bmatrix}, \quad \mathbf{q} = \begin{bmatrix} q_1 \\ q_2 \end{bmatrix}. \tag{4.7}$$

The function vector $\mathbf{f}$ and its reference are now given by

$$\mathbf{f} = \dot{\boldsymbol{\varphi}} \tag{4.8}$$

$$\mathbf{f}^{ref} = \dot{\boldsymbol{\varphi}} - \mathbf{C} \left(\boldsymbol{\varphi} - \boldsymbol{\varphi}^{ref} \right), \quad \mathbf{C} = \begin{bmatrix} 10 & 0 \\ 0 & 10 \end{bmatrix}. \tag{4.9}$$

The differential equation $\mathbf{f}(\mathbf{q}, \mathbf{v}) = \mathbf{f}^{ref}(t)$ does not have a solution in the set of real numbers, so the control goal cannot be achieved. One can confirm this just by noticing that it is not possible to find real q_1 and q_2 so that $\boldsymbol{\varphi}(\mathbf{q}) = \boldsymbol{\varphi}^{ref}(t)$ is satisfied. Therefore, the control goal given in (4.1) is not feasible. On the other hand, for example, for $\boldsymbol{\varphi}^{ref}(t) = [4\ 2]^T$ there are two possible configuration vectors which satisfy $\boldsymbol{\varphi}(\mathbf{q}) = \boldsymbol{\varphi}^{ref}(t)$, given as $\mathbf{q}_{f1} = [0\ 2]^T$ and $\mathbf{q}_{f2} = [2\ 0]^T$. However, for $\boldsymbol{\varphi}^{ref}(t) = [4\ 2\sqrt{2}]^T$, the equation $\boldsymbol{\varphi}(\mathbf{q}) = \boldsymbol{\varphi}^{ref}(t)$ has one and only one solution $\mathbf{q}_f = [\sqrt{2}\ \sqrt{2}]^T$. Therefore, the selection of functions and their references has to be done carefully.

Following the notation introduced in the previous chapters, the first-order dynamics of the function vector is

$$\dot{\mathbf{f}} = \mathbf{J}_f(\mathbf{q}, \mathbf{v})\,\dot{\mathbf{v}} + \boldsymbol{\Upsilon} = \mathbf{J}_f(\mathbf{q}, \mathbf{v})\,\ddot{\mathbf{q}} + \boldsymbol{\Upsilon}. \tag{4.10}$$

where the function Jacobian $\mathbf{J}_f \in \mathbb{R}^{m \times n}$ is defined as explained in Section 3.2.1.

When the functions are being selected, the main goal is to be able to control all functions. In other words, the input force vector in the configuration space has to be able to enforce desired dynamics of all functions. This is equivalent to enforcing each function to track its corresponding reference. The input force vector controls the acceleration vector in the configuration space $\dot{\mathbf{v}} \in \mathbb{R}^{n \times 1}$. Therefore, in order to control dynamics of the functions one can find desired acceleration in the configuration space $\dot{\mathbf{v}}^{des} \in \mathbb{R}^{n \times 1}$. After that, the input force vector can be selected to enforce

$$\dot{\mathbf{v}} = \dot{\mathbf{v}}^{des}. \tag{4.11}$$

This was discussed in Section 3.5.1. The question to be answered is whether $\dot{\mathbf{v}}^{des}$ which enforces the desired dynamics of all functions can be found at all. As already stated in Section 3.5.1, this can be done if and only if $\mathbf{J}_f$ is a full row rank matrix, which follows directly from the Kronecker–Capelli theorem.

From the already given discussion, it follows that functions can be controlled as long as $\mathbf{J}_f$ has full row rank. Since it is $m \leq n$, full row rank of $\mathbf{J}_f$ actually means full rank of the matrix. Therefore, it is very important to monitor the rank during the system operation. The simplest case is when $\mathbf{J}_f$ is a constant matrix. Its rank can be then calculated before the system starts to operate, or in the moments when function set is being changed. In other cases, the rank of $\mathbf{J}_f$ should be monitored all the time since in general it depends on the configuration vector $\mathbf{q}$ and configuration space velocity vector $\mathbf{v}$.

Monitoring of the $\mathbf{J}_f$ rank can be done in different ways. If the matrix is a square matrix its determinant can be used as an indication of whether $\mathbf{J}_f$ has full rank. Otherwise, in general, one can look at the number of nonzero singular values of $\mathbf{J}_f$ which will be denoted as r. The rank of the matrix $\mathbf{J}_f$ is then equal to r. The number r can be obtained from singular value decomposition of the matrix $\mathbf{J}_f$. On the other hand, the positive square roots of the nonzero eigenvalues of the positive semidefinite square Gram matrix $\mathbf{Q} = \mathbf{J}_f^T \mathbf{J}_f$ are equal to nonzero singular values of $\mathbf{J}_f$. Therefore, r equals the number of these positive square roots.

It is important to consider in more detail a case when $\mathbf{f}(\mathbf{q}, \mathbf{v}) = \mathbf{f}^{ref}(t)$ has multiple solutions, which was shown to be possible in an example given above. For the vector $\boldsymbol{\varphi}(\mathbf{q})$ from (4.7) and $\boldsymbol{\varphi}^{ref}(t) = [4 \ 2]^T$, two possible solutions of the equation $\mathbf{f}(\mathbf{q}, \mathbf{v}) = \mathbf{f}^{ref}(t)$ are $\mathbf{q}_{f1} = [0 \ 2]^T$ and $\mathbf{q}_{f2} = [2 \ 0]^T$, as already said. The main question now is which of these two solutions will be achieved after the control system enforces convergence to the reference. To answer this question, it will be assumed that the perfect control in the configuration space is achieved, meaning that (4.11) holds. The vector $\dot{\mathbf{v}}^{des}$ is calculated as the solution $\dot{\mathbf{v}}$ of (3.132) as

$$\dot{\mathbf{v}}^{des} = \mathbf{J}_f^{-1}\left(\dot{\mathbf{f}}^{des} - \boldsymbol{\Upsilon}\right), \quad \boldsymbol{\Upsilon} = \dot{\mathbf{J}}_f \mathbf{v} \tag{4.12}$$

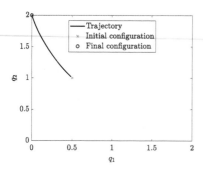
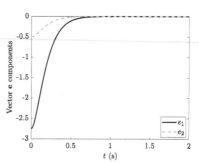

FIGURE 4.1

System trajectory in the first case.

FIGURE 4.2

Vector **e** components in the first case.

assuming that $\mathbf{J}_f$ is a nonsingular matrix, while $\dot{\mathbf{f}}^{des}$ is given by

$$\dot{\mathbf{f}}^{des} = \dot{\mathbf{f}}^{ref} - 10\left(\mathbf{f} - \mathbf{f}^{ref}\right), \quad \mathbf{f}^{ref} = \dot{\boldsymbol{\varphi}}^{ref} - 10\mathbf{e}, \quad \mathbf{e} = [e_1 \ e_2]^{\mathrm{T}} = \boldsymbol{\varphi} - \boldsymbol{\varphi}^{ref}. \tag{4.13}$$

First, the initial conditions were taken as

$$\mathbf{q}(0) = \begin{bmatrix} 0.5 \\ 1 \end{bmatrix}, \dot{\mathbf{q}}(0) = \begin{bmatrix} 0 \\ 0 \end{bmatrix}. \tag{4.14}$$

Responses are shown in Figures 4.1 and 4.2. The controlled system is converging to the solution $\mathbf{q}_{f1}$.

In the second case, the initial conditions were

$$\mathbf{q}(0) = \begin{bmatrix} 1.5 \\ 1 \end{bmatrix}, \dot{\mathbf{q}}(0) = \begin{bmatrix} 0 \\ 0 \end{bmatrix}. \tag{4.15}$$

Responses are shown in Figures 4.3 and 4.4. The controlled system is converging to the solution $\mathbf{q}_{f2}$. To answer the posed question, one can say the following. To which of the two solutions, $\mathbf{q}_{f1}$ or $\mathbf{q}_{f2}$, will the system converge actually depends on the initial configuration of the system. The system is controlled to enforce exponential decay of the vector **e** components. Therefore, from its initial configuration, the system is converging to the solution that assures such behavior of the vector **e** components. In the first case, this can be achieved only if the system is converging to $\mathbf{q}_{f1}$, while in the second case, the desired behavior is enforced when the system is converging to $\mathbf{q}_{f2}$. An alternative explanation can be made. The target configuration is $\mathbf{q}_{f1}$ or $\mathbf{q}_{f2}$. In the first case, the initial configuration is closer to $\mathbf{q}_{f1}$ and the system is converging to $\mathbf{q}_{f1}$. Similarly, in the second case, the initial configuration is closer to $\mathbf{q}_{f2}$; thus, the system is converging to $\mathbf{q}_{f2}$.

Besides the mentioned constraints on the selection of the function vector and its reference, it is important to mention that both $\mathbf{f}$ and $\mathbf{f}^{ref}$ have

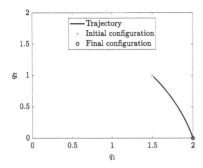

FIGURE 4.3

System trajectory in the second case.

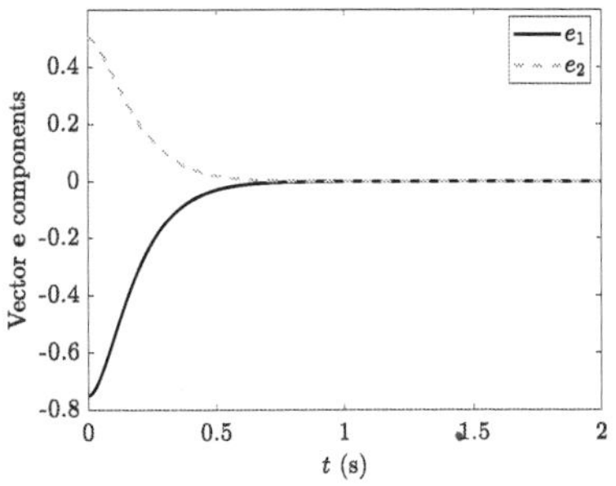

FIGURE 4.4

Vector **e** components in the second case.

to be differentiable with respect to time. It has to be emphasized that further research regarding the selection of functions is still needed, in order to formalize this process.

4.2 Closed-Loop Dynamics in Configuration Space

The approach presented in Section 3.5.1 is basically equivalent to the control design approaches presented in Section 3.5.2, since they all enforce some desired dynamics of the vector **f**. However, there is a detail that has to be considered once the dynamics of the closed-loop system in the configuration space is analyzed for $m < n$, and it will be presented here. For the sake of comparison, the third approach for control synthesis in the function space, elaborated in Section 3.5.2.3, will be compared with the approach presented in Section 3.5.1. The system is described in the form (3.6). The tracking error vector $\mathbf{e} \in \mathbb{R}^{m \times 1}$ and generalized error $\boldsymbol{\sigma} \in \mathbb{R}^{m \times 1}$ are defined as

$$
\begin{aligned}
\mathbf{e} &= \boldsymbol{\varphi} - \boldsymbol{\varphi}^{ref} \\
\boldsymbol{\sigma} &= \mathbf{f} - \mathbf{f}^{ref}.
\end{aligned}
\tag{4.16}
$$

The control goal will be fulfilled if the system motion is converging to the manifold $\boldsymbol{\sigma} = \mathbf{0}$. Considering (3.4), (4.10), and (4.16), the first-order dynamics of the generalized error is

$$
\dot{\boldsymbol{\sigma}} = \mathbf{J}_f \mathbf{u}_q - \left[\mathbf{J}_f \mathbf{A}^{-1} \left(\mathbf{b} + \mathbf{g} + \mathbf{T}_{ext} \right) - \boldsymbol{\Upsilon} + \dot{\mathbf{f}}^{ref} \right].
\tag{4.17}
$$

For $m < n$, $\mathbf{u}_q$ is expressed as $\mathbf{u}_q = \mathbf{J}_f^{\#} \mathbf{u}_f + \boldsymbol{\Gamma} \mathbf{u}_{q_0}$, $\mathbf{I} - \mathbf{J}_f^{\#} \mathbf{J}_f = \boldsymbol{\Gamma} \in \mathbb{R}^{n \times n}$, where $\mathbf{u}_{q_0}$ is an arbitrary acceleration vector in $\mathbb{R}^{n \times 1}$. One can notice

$\mathbf{J}_f \mathbf{u}_q = \mathbf{u}_f \in \mathbb{R}^{m \times 1}$. By introducing the equivalent control as $\mathbf{J}_f \mathbf{A}^{-1}$
$(\mathbf{b} + \mathbf{g} + \mathbf{T}_{ext}) - \mathbf{\Upsilon} + \dot{\mathbf{f}}^{ref} = \mathbf{u}_f^{eq} \in \mathbb{R}^{m \times 1}$, dynamics (4.17) becomes

$$\dot{\boldsymbol{\sigma}} = \mathbf{u}_f - \mathbf{u}_f^{eq}. \tag{4.18}$$

It will be assumed that the perfect equivalent control estimation is possible, and $\mathbf{u}_f^{eq}$ is available. The control vector in the function space is selected as

$$\mathbf{u}_f = \mathbf{u}_f^{eq} - \mathbf{\Psi}(\boldsymbol{\sigma}) = \mathbf{J}_f \mathbf{A}^{-1}(\mathbf{b} + \mathbf{g} + \mathbf{T}_{ext}) - \mathbf{\Upsilon} + \dot{\mathbf{f}}^{ref} - \mathbf{\Psi}(\boldsymbol{\sigma}) \tag{4.19}$$

where $\mathbf{\Psi}(\boldsymbol{\sigma}) \in \mathbb{R}^{m \times 1}$ is a function selected such that $\dot{\boldsymbol{\sigma}} + \mathbf{\Psi}(\boldsymbol{\sigma}) = \mathbf{0}$ defines a desired dynamics of the generalized error $\boldsymbol{\sigma}$. The desired dynamics has to ensure $\boldsymbol{\sigma} \xrightarrow{t \to \infty} \mathbf{0}$. For this control, the closed-loop dynamics of the generalized error is

$$\dot{\boldsymbol{\sigma}} + \mathbf{\Psi}(\boldsymbol{\sigma}) = \mathbf{0}. \tag{4.20}$$

The control acceleration is

$$\mathbf{u}_q = \mathbf{J}_f^{\#}\left[\mathbf{J}_f \mathbf{A}^{-1}(\mathbf{b} + \mathbf{g} + \mathbf{T}_{ext}) - \mathbf{\Upsilon} + \dot{\mathbf{f}}^{ref} - \mathbf{\Psi}(\boldsymbol{\sigma})\right] + \mathbf{\Gamma}\mathbf{u}_{q_0}. \tag{4.21}$$

and the input force $\mathbf{T} = \mathbf{A}\mathbf{u}_q$ is

$$\mathbf{T} = \mathbf{A}\mathbf{J}_f^{\#}\left[\mathbf{J}_f \mathbf{A}^{-1}(\mathbf{b} + \mathbf{g} + \mathbf{T}_{ext}) - \mathbf{\Upsilon} + \dot{\mathbf{f}}^{ref} - \mathbf{\Psi}(\boldsymbol{\sigma})\right] + \mathbf{A}\mathbf{\Gamma}\mathbf{u}_{q_0}. \tag{4.22}$$

When $\mathbf{u}_q$ from (4.21) is used in (3.6), considering also the closed-loop dynamics of the generalized error given in (4.20), the total closed-loop dynamics of the system is

$$\left.\begin{array}{r}\dot{\mathbf{q}} = \mathbf{v} \\ \dot{\mathbf{v}} + \left(\mathbf{I} - \mathbf{J}_f^{\#}\mathbf{J}_f\right)\mathbf{A}^{-1}(\mathbf{b} + \mathbf{g} + \mathbf{T}_{ext}) = \mathbf{J}_f^{\#}\left[\dot{\mathbf{f}}^{ref} - \mathbf{\Psi}(\boldsymbol{\sigma})\right] - \mathbf{J}_f^{\#}\mathbf{\Upsilon} + \mathbf{\Gamma}\mathbf{u}_{q_0} \\ \dot{\boldsymbol{\sigma}} + \mathbf{\Psi}(\boldsymbol{\sigma}) = \mathbf{0}.\end{array}\right\} \tag{4.23}$$

The same form can be obtained if $\mathbf{T}$ from (4.22) is used in (2.1), taking into account the fact that $\mathbf{A}$ is a nonsingular matrix. If the term $\dot{\mathbf{f}}^{ref} - \mathbf{\Psi}(\boldsymbol{\sigma})$ is denoted as $\dot{\mathbf{f}}^{des}$, since $\dot{\mathbf{f}} = \dot{\mathbf{f}}^{des}$ is a different form for (4.20), (4.23) can also be written as

$$\left.\begin{array}{r}\dot{\mathbf{q}} = \mathbf{v} \\ \dot{\mathbf{v}} + \left(\mathbf{I} - \mathbf{J}_f^{\#}\mathbf{J}_f\right)\mathbf{A}^{-1}(\mathbf{b} + \mathbf{g} + \mathbf{T}_{ext}) = \mathbf{J}_f^{\#}\dot{\mathbf{f}}^{des} - \mathbf{J}_f^{\#}\mathbf{\Upsilon} + \mathbf{\Gamma}\mathbf{u}_{q_0} \\ \dot{\mathbf{f}} = \dot{\mathbf{f}}^{des}.\end{array}\right\} \tag{4.24}$$

For the case when the desired acceleration in the configuration space is enforced, the desired acceleration is calculated as $\dot{\mathbf{v}}^{des} = \mathbf{J}_f^{\#}\left(\dot{\mathbf{f}}^{des} - \mathbf{\Upsilon}\right) + \mathbf{\Gamma}\ddot{\mathbf{q}}_0$ where $\ddot{\mathbf{q}}_0$ is an arbitrary acceleration vector in $\mathbb{R}^{n \times 1}$. The desired acceleration is enforcing

$$\dot{\mathbf{f}} = \dot{\mathbf{f}}^{des}. \tag{4.25}$$

where $\dot{\mathbf{f}}^{des}$ is the desired dynamics of the function vector. For simplicity, the perfect disturbance estimation in the configuration space is assumed, i.e., $\mathbf{u}_{qdis}$ from (3.8) is available. The control acceleration is selected as

$$\mathbf{u}_q = \mathbf{u}_{qdis} + \dot{\mathbf{v}}^{des} = \mathbf{A}^{-1}\left(\mathbf{b} + \mathbf{g} + \mathbf{T}_{ext}\right) + \mathbf{J}_f^{\#}\left(\dot{\mathbf{f}}^{des} - \mathbf{\Upsilon}\right) + \mathbf{\Gamma}\ddot{\mathbf{q}}_0 \quad (4.26)$$

and the input force $\mathbf{T} = \mathbf{A}\mathbf{u}_q$ is

$$\mathbf{T} = \mathbf{b} + \mathbf{g} + \mathbf{T}_{ext} + \mathbf{A}\mathbf{J}_f^{\#}\left(\dot{\mathbf{f}}^{des} - \mathbf{\Upsilon}\right) + \mathbf{A}\mathbf{\Gamma}\ddot{\mathbf{q}}_0. \quad (4.27)$$

When $\mathbf{u}_q$ from (4.26) is used in (3.6), considering also the dynamics of the function vector given in (4.25), the total closed-loop dynamics of the system is

$$\left.\begin{array}{c}\dot{\mathbf{q}} = \mathbf{v} \\ \dot{\mathbf{v}} = \mathbf{J}_f^{\#}\dot{\mathbf{f}}^{des} - \mathbf{J}_f^{\#}\mathbf{\Upsilon} + \mathbf{\Gamma}\ddot{\mathbf{q}}_0 \\ \dot{\mathbf{f}} = \dot{\mathbf{f}}^{des}.\end{array}\right\} \quad (4.28)$$

The same form can be obtained if $\mathbf{T}$ from (4.27) is used in (2.1), taking into account the fact that $\mathbf{A}$ is a nonsingular matrix. If (4.24) and (4.28) are compared one can notice the following; for $\mathbf{u}_{q_0} = \ddot{\mathbf{q}}_0$, the only difference would be that (4.24) contains the term $\left(\mathbf{I} - \mathbf{J}_f^{\#}\mathbf{J}_f\right)\mathbf{A}^{-1}\left(\mathbf{b} + \mathbf{g} + \mathbf{T}_{ext}\right) = \mathbf{\Gamma}\mathbf{A}^{-1}\left(\mathbf{b} + \mathbf{g} + \mathbf{T}_{ext}\right)$. This term represents the acceleration induced by the force $\mathbf{b} + \mathbf{g} + \mathbf{T}_{ext}$ projected by the matrix $\mathbf{\Gamma}$, which is the null space projection matrix associated with the function Jacobian $\mathbf{J}_f$. This term appears since the force $\mathbf{b} + \mathbf{g} + \mathbf{T}_{ext}$ is not compensated in the configuration space, which is the case for the method based on enforcement of the desired acceleration.

In the case when the number of the functions is equal to the dimension of the configuration vector $\mathbf{q}$, i.e., when $m = n$, both analyzed methods will induce the same closed-loop dynamics described by

$$\mathbf{J}_f\dot{\mathbf{v}} + \mathbf{\Upsilon} = \dot{\mathbf{f}}^{des} \Leftrightarrow \dot{\mathbf{f}} = \dot{\mathbf{f}}^{des}. \quad (4.29)$$

4.3 Zero Dynamics

As was already stated, the control goal is to enforce (3.128) and consequently (4.1). It is important to analyze the closed-loop dynamics of the system when the control goal is achieved, i.e., when the stationary state is reached, which is considered as the zero dynamics of the system. Thus, for the case $m < n$, the analysis will show which form (4.23) will have. Due to the third equation in (4.23) that ensures $\boldsymbol{\sigma} \xrightarrow{t \to \infty} \mathbf{0}$ one can assume that after sufficient time the system motion is stable in the manifold $\boldsymbol{\sigma} = \mathbf{f} - \mathbf{f}^{ref} = \mathbf{0}$. After the control goal is achieved, the control vector in the function space becomes $\mathbf{u}_f = \mathbf{u}_f^{eq}$.

For such $\mathbf{u}_f$, using the same procedure as is applied for obtaining the second equation in (4.23), the second equation in (4.23) becomes

$$\dot{\mathbf{v}} + \left(\mathbf{I} - \mathbf{J}_f^{\#}\mathbf{J}_f\right)\mathbf{A}^{-1}\left(\mathbf{b} + \mathbf{g} + \mathbf{T}_{ext}\right) = \mathbf{J}_f^{\#}\dot{\mathbf{f}}^{ref} - \mathbf{J}_f^{\#}\Upsilon + \Gamma\mathbf{u}_{q_0}. \qquad (4.30)$$

The third equation in (4.23), which describes behavior of the function vector, becomes the differential equation

$$\mathbf{f}\left(\mathbf{q}, \mathbf{v}\right) = \mathbf{f}^{ref}\left(t\right) \qquad (4.31)$$

when the control goal is achieved. Therefore, the total closed-loop dynamics is now

$$\left. \begin{array}{c} \dot{\mathbf{q}} = \mathbf{v} \\ \dot{\mathbf{v}} + \left(\mathbf{I} - \mathbf{J}_f^{\#}\mathbf{J}_f\right)\mathbf{A}^{-1}\left(\mathbf{b} + \mathbf{g} + \mathbf{T}_{ext}\right) = \mathbf{J}_f^{\#}\dot{\mathbf{f}}^{ref} - \mathbf{J}_f^{\#}\Upsilon + \Gamma\mathbf{u}_{q_0} \\ \mathbf{f}\left(\mathbf{q}, \mathbf{v}\right) = \mathbf{f}^{ref}\left(t\right). \end{array} \right\} \qquad (4.32)$$

The third equation in (4.32) represents m first-order differential equations, where the unknowns are the components of the configuration vector $\mathbf{q}$. If $\mathbf{f}\left(\mathbf{q}, \mathbf{v}\right)$ and $\mathbf{f}^{ref}$ are properly selected, one can find the solution for m components of the vector $\mathbf{q}$. These m components will be expressed as functions of the remaining $(n - m)$ components of the vector $\mathbf{q}$ and corresponding $(n - m)$ components of the vector $\mathbf{v}$. When these m components of the configuration vector are used in the second equation of (4.32), the system dynamics becomes $(n - m)$-dimensional. For $m = n$, the system is fully described by the first and third equation in (4.32), i.e., the closed-loop dynamics is given by

$$\left. \begin{array}{c} \dot{\mathbf{q}} = \mathbf{v} \\ \mathbf{f}\left(\mathbf{q}, \mathbf{v}\right) = \mathbf{f}^{ref}\left(t\right). \end{array} \right\} \qquad (4.33)$$

4.4   Conclusion

In this chapter, constraints and conditions to be considered when specifying functions to be controlled in a motion control system have been discussed. It has been shown that functions and their references have to be carefully selected so the functions can be controlled. If references and vectors are not properly selected, it may happen that functions cannot ever be made to converge to the references. For investigating whether the functions can be controlled at each instance of time, one needs to monitor the rank of the Jacobian matrix. In addition, in this chapter, the closed-loop dynamics of the controlled system in the configuration space has been discussed. It has been done for control methods from the previous chapter.

5

Motion Synchronization and Object Manipulation in 2-D Space

The method proposed in Chapter 3 for control of functionally related systems is experimentally applied for two different tasks and these experiments are discussed in this chapter. The executed tasks are: (i) motion synchronization in two-dimensional (2-D) space for two pantograph manipulators and (ii) object manipulation in 2-D space which again included these two manipulators. Experimental results demonstrate the effectiveness of the presented method.

5.1 Experimental Setup

The experimental setup used for experimental validation of the proposed approach to design of a motion control system for functionally related systems is shown in Figure 5.1. Two main parts of the setup are two pantograph manipulators. CAD design of one manipulator with its overall dimensions is given in Figure 5.2. A photo of one of the manufactured manipulators is depicted in Figure 5.3. The manipulators can be aligned and connected with a bar, which is used in the experiments that included object manipulation. The manipulators are actuated with Faulhaber LM 2070-080-11 linear motors. The motors are driven by Faulhaber MCLM 3006 S RS motor drivers. Displacements of the motors are measured by Renishaw RGH41X30D05A encoders which provide resolution of 1 μm. The manipulators are controlled by a modular real-time control system produced by dSPACE. The modular system is based on a DS1005 processor board, and it also has a DS3001 board used for interfacing encoders and a DS2103 D/A board which is utilized as an interface to the motor drivers. The dSPACE system is connected with a supervisory computer to provide the possibility for real time recording of different signals, monitoring, and supervisory control of the real-time process being executed in the control system. The supervisory computer is an HP xw6400 Workstation which has a dual-core processor Intel Xeon E5345 @ 2.33 GHz, 3 GB of RAM, and 32-bit Windows 7 operating system.

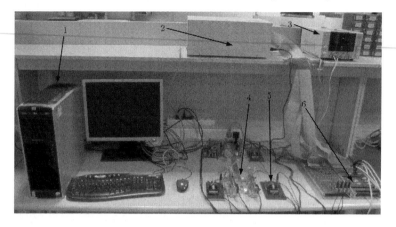

FIGURE 5.1
Experimental setup. 1 - supervisory computer, 2 - dSPACE system, 3 - power supply, 4 - one of two pantograph manipulators, 5 - one of four motor drivers, 6 - connection boards for dSPACE system.

5.2   Direct Kinematics of Pantograph Manipulator

The experiments presented in this chapter involved two identical manipulators, and the design of a manipulator is illustrated in Figure 5.4. The manipulator is actuated by two linear motors, which are connected to two links of length l by two rotational bearings. The length l is measured from the center of a bearing to the center of the end-effector. The displacements of the motors are q_1 and q_2, and they form the configuration vector of the manipulator $\mathbf{q} = [q_1 \ q_2]^\mathrm{T}$. The motors are at the distance $2h$ from each other. In order to derive the direct kinematics model for the pantograph manipulator, the origin of the inertial coordinate system is positioned at the line connecting centers of the rotational bearings, when both motors are displaced for zero amount, i.e., when $q_1 = q_2 = 0$, and it is on the center of this line.

The position of the end-effector (its center to be specific) in the inertial frame (task space) is given as $\mathbf{x} = [x \ y]^\mathrm{T}$. When the motors move for q_1 and q_2, since the distance between the centers of the bearings and end-effector is known to be equal to the link length l, the following equations can be written

$$
\begin{aligned}
(x - q_1)^2 + (y - h)^2 &= l^2 \\
(x - q_2)^2 + (y + h)^2 &= l^2
\end{aligned}
\tag{5.1}
$$

which is equivalent to

$$
\begin{aligned}
x^2 - 2xq_1 + q_1^2 + y^2 - 2hy + h^2 &= l^2 \\
x^2 - 2xq_2 + q_2^2 + y^2 + 2hy + h^2 &= l^2.
\end{aligned}
\tag{5.2}
$$

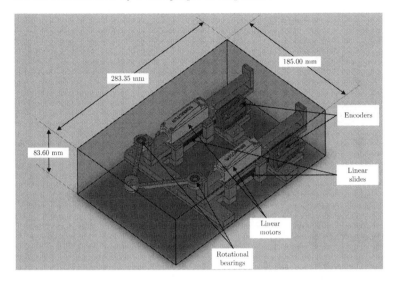

FIGURE 5.2
CAD design and overall dimensions of pantograph manipulator.

When one subtracts the second equation from the first equation in (5.2), y can be expressed as a function of x as

$$y = \underbrace{\left(\frac{q_2 - q_1}{2h}\right)}_{\alpha} x + \underbrace{\left(\frac{q_1^2 - q_2^2}{4h}\right)}_{\beta} = \alpha x + \beta. \qquad (5.3)$$

Substituting (5.3) in the second equation in (5.2), it can be written

$$\underbrace{\left(\alpha^2 + 1\right)}_{a} x^2 + \underbrace{\left(2\alpha\beta + 2h\alpha - 2q_2\right)}_{b} x + \underbrace{\left(q_2^2 + \beta^2 + 2h\beta + h^2 - l^2\right)}_{c} = 0. \qquad (5.4)$$

The last equation is a quadratic equation with roots given as

$$x_1 = \frac{-b - \sqrt{\Delta}}{2a}, \quad x_2 = \frac{-b + \sqrt{\Delta}}{2a} \qquad (5.5)$$

where

$$\Delta = b^2 - 4ac. \qquad (5.6)$$

Since $\Delta > 0$ (assuming no singular configuration) there are two solutions for the x-coordinate, where $x_2 > x_1$. Due to the configuration of the designed pantograph manipulator, the right solution is x_2. Therefore, the final expressions for x and y are as follows

$$x = \frac{-b + \sqrt{b^2 - 4ac}}{2a} \qquad (5.7)$$

$$y = \alpha x + \beta \qquad (5.8)$$

FIGURE 5.3
Photo of one pantograph manipulator.

where the coefficients appearing in (5.7) and (5.8) are

$$\alpha = \left(\frac{q_2 - q_1}{2h}\right) \tag{5.9}$$

$$\beta = \left(\frac{q_1^2 - q_2^2}{4h}\right) \tag{5.10}$$

$$a = \left(\alpha^2 + 1\right) \tag{5.11}$$

$$b = (2\alpha\beta + 2h\alpha - 2q_2) \tag{5.12}$$

$$c = \left(q_2^2 + \beta^2 + 2h\beta + h^2 - l^2\right). \tag{5.13}$$

For deriving the velocity level kinematics, one can differentiate (5.2) with respect to time and get

$$\begin{aligned}
(x - q_1)\,\dot{x} + (y - h)\,\dot{y} &= (x - q_1)\,\dot{q}_1 \\
(x - q_2)\,\dot{x} + (y + h)\,\dot{y} &= (x - q_2)\,\dot{q}_2.
\end{aligned} \tag{5.14}$$

Assuming $x - q_1 \neq 0$, since $x - q_1 = 0$ is possible only for a constrained mechanism which has one degree of freedom and then it is $x = q_1 = q_2$, it can be written

$$\begin{bmatrix} \dot{q}_1 \\ \dot{q}_2 \end{bmatrix} = \underbrace{\begin{bmatrix} 1 & \frac{y-h}{x-q_1} \\ 1 & \frac{y+h}{x-q_2} \end{bmatrix}}_{\mathbf{J}^{-1}} \begin{bmatrix} \dot{x} \\ \dot{y} \end{bmatrix} = \mathbf{J}^{-1} \begin{bmatrix} \dot{x} \\ \dot{y} \end{bmatrix} \tag{5.15}$$

where $\mathbf{J}^{-1}$ is the inverse Jacobian matrix for the manipulator.

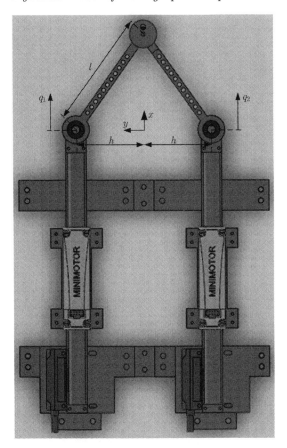

FIGURE 5.4
Pantograph manipulator with the attached inertial frame.

5.3 Dynamic Model of Pantograph Manipulator

In this chapter, it will be considered that a single pantograph manipulator can be modeled in the configuration space as follows

$$\left. \begin{array}{l} \dot{\mathbf{q}} = \mathbf{v} \\ \mathbf{A}\left(\mathbf{q}\right) \dot{\mathbf{v}} = \mathbf{T} - \mathbf{T}_l\left(t\right) \end{array} \right\} \tag{5.16}$$

where $\mathbf{q} \in \mathbb{R}^{2 \times 1}$ is the configuration vector, $\mathbf{v} \in \mathbb{R}^{2 \times 1}$ is the configuration space velocity vector, $\mathbf{A}\left(\mathbf{q}\right) \in \mathbb{R}^{2 \times 2}$ represents the symmetric positive definite kinetic energy matrix (inertia matrix), $\mathbf{T} \in \mathbb{R}^{2 \times 1}$ denotes the vector of control forces generated by the motors and it will be denoted as the input force vector in the configuration space, and $\mathbf{T}_l\left(t\right) \in \mathbb{R}^{2 \times 1}$ represents all other forces acting on the motion of manipulator motors, like Coriolis forces, viscous friction forces, centripetal forces, and additional external forces influencing the manipulator. If the matrix $\mathbf{A}$ is written as a sum of its nominal value, that is a constant symmetric positive definite matrix, and variation, i.e., as $\mathbf{A}(\mathbf{q}) = \mathbf{A}_n + \Delta\mathbf{A}(\mathbf{q})$, dynamics (5.16) can be rewritten as

$$\mathbf{A}_n\ddot{\mathbf{q}} = \mathbf{T} - \underbrace{(\mathbf{T}_l + \Delta\mathbf{A}\ddot{\mathbf{q}})}_{\mathbf{T}_{dis}} = \mathbf{T} - \mathbf{T}_{dis}. \qquad (5.17)$$

where $\mathbf{T}_{dis} \in \mathbb{R}^{2\times1}$ is the generalized disturbance. The last equation will be used in control derivation for the two tasks.

5.4   Motion Synchronization Task in 2-D Space

The goal in the first task was to make positions of the pantograph manipulators (positions of their end-effectors, to be more precise) equal in the task space, and to have them track a reference trajectory. It is assumed that each manipulator's task space position is expressed in the x-y frame attached to that manipulator as explained in Section 5.2. Let us assume that positions of the manipulators in their corresponding frames are given as $\mathbf{x}_1(\mathbf{q}_1) = [x_1 \; y_1]^{\mathrm{T}}$ and $\mathbf{x}_2(\mathbf{q}_2) = [x_2 \; y_2]^{\mathrm{T}}$, where $\mathbf{q}_1$ and $\mathbf{q}_2$ stand for the configuration vectors of the first and second manipulator, respectively. The reference trajectory is given as a two times differentiable vector-valued function of time $\mathbf{x}^{ref}(t) = [x^{ref}(t) \; y^{ref}(t)]^{\mathrm{T}}$. For the task to be executed, two functions should be controlled, namely: (i) synchronization function $\boldsymbol{\varphi}_s = [\varphi_{s1} \; \varphi_{s2}]^{\mathrm{T}}$, and (ii) reference tracking function $\boldsymbol{\varphi}_{rt} = [\varphi_{rt1} \; \varphi_{rt2}]^{\mathrm{T}}$, where each of them is a vector valued function, i.e., $\boldsymbol{\varphi}_s, \boldsymbol{\varphi}_{rt} \in \mathbb{R}^{2\times1}$. These functions can be expressed in terms of manipulators' positions as

$$\boldsymbol{\varphi}(\mathbf{q}_1, \mathbf{q}_2) = \begin{bmatrix} \boldsymbol{\varphi}_s(\mathbf{q}_1, \mathbf{q}_2) \\ \boldsymbol{\varphi}_{rt}(\mathbf{q}_1, \mathbf{q}_2) \end{bmatrix} = \begin{bmatrix} \mathbf{x}_1 - \mathbf{x}_2 \\ \mathbf{x}_1 + \mathbf{x}_2 \end{bmatrix}. \qquad (5.18)$$

The task will be executed if both of these functions track their references $\boldsymbol{\varphi}_s^{ref}(t)$ and $\boldsymbol{\varphi}_{rt}^{ref}(t)$ defined as

$$\boldsymbol{\varphi}^{ref}(t) = \begin{bmatrix} \boldsymbol{\varphi}_s^{ref}(t) \\ \boldsymbol{\varphi}_{rt}^{ref}(t) \end{bmatrix} = \begin{bmatrix} \mathbf{0} \\ 2\mathbf{x}^{ref} \end{bmatrix}. \qquad (5.19)$$

When $\boldsymbol{\varphi}_s = \boldsymbol{\varphi}_s^{ref}$ is enforced, then $\mathbf{x}_1 = \mathbf{x}_2$. Additional enforcement of $\boldsymbol{\varphi}_{rt} = \boldsymbol{\varphi}_{rt}^{ref}$ ensures that $\mathbf{x}_1 = \mathbf{x}_2 = \mathbf{x}^{ref}$, which means that the task is executed. The defined functions, $\boldsymbol{\varphi}_s$ and $\boldsymbol{\varphi}_{rt}$ depend on the configuration vectors $\mathbf{q}_1$ and $\mathbf{q}_2$ only. Thus, one can form the function vector, following the procedure given in Section 3.2.1, as follows

$$\mathbf{f} = \dot{\boldsymbol{\varphi}}. \qquad (5.20)$$

According to the procedure given in Section 4.1, the function vector reference is

$$\mathbf{f}^{ref} = \dot{\boldsymbol{\varphi}}^{ref} - \mathbf{C}\left(\boldsymbol{\varphi} - \boldsymbol{\varphi}^{ref}\right) \tag{5.21}$$

where $\mathbf{C} \in \mathbb{R}^{4\times 4}$ is a constant diagonal matrix with positive diagonal entries

$$\mathbf{C} = \text{diag}\left(c_1,\ c_2,\ c_3,\ c_4\right),\ c_i > 0,\ i = 1,\ 2,\ 3,\ 4. \tag{5.22}$$

The dynamics of the two manipulators can be written as

$$\left.\begin{aligned} \dot{\mathbf{q}}_1 &= \mathbf{v}_1 \\ \mathbf{A}_{1n}\dot{\mathbf{v}}_1 &= \mathbf{T}_1 - \mathbf{T}_{1dis} \end{aligned}\right\} \tag{5.23}$$

$$\left.\begin{aligned} \dot{\mathbf{q}}_2 &= \mathbf{v}_2 \\ \mathbf{A}_{2n}\dot{\mathbf{v}}_2 &= \mathbf{T}_2 - \mathbf{T}_{2dis} \end{aligned}\right\} \tag{5.24}$$

where subscripts 1 and 2 refer to the first and second manipulator. Components of the input force vectors in the configuration space, $\mathbf{T}_1$ and $\mathbf{T}_2$, and generalized disturbances, $\mathbf{T}_{1dis}$ and $\mathbf{T}_{2dis}$, are given as

$$\mathbf{T}_1 = \begin{bmatrix} T_{11} \\ T_{21} \end{bmatrix},\ \mathbf{T}_2 = \begin{bmatrix} T_{12} \\ T_{22} \end{bmatrix},\ \mathbf{T}_{1dis} = \begin{bmatrix} T_{11dis} \\ T_{21dis} \end{bmatrix},\ \mathbf{T}_{2dis} = \begin{bmatrix} T_{12dis} \\ T_{22dis} \end{bmatrix} \tag{5.25}$$

In a shorter form, the dynamics of the system that consists of both manipulators can be written as

$$\left.\begin{aligned} \dot{\tilde{\mathbf{q}}} &= \tilde{\mathbf{v}} \\ \tilde{\mathbf{A}}_n\dot{\tilde{\mathbf{v}}} &= \tilde{\mathbf{T}} - \tilde{\mathbf{T}}_{dis}. \end{aligned}\right\} \tag{5.26}$$

The matrices and vectors that appear in (5.26) are given as

$$\tilde{\mathbf{A}}_n = \begin{bmatrix} \mathbf{A}_{1n} & \mathbf{0}^{2\times 2} \\ \mathbf{0}^{2\times 2} & \mathbf{A}_{2n} \end{bmatrix},\ \tilde{\mathbf{q}} = \begin{bmatrix} \mathbf{q}_1 \\ \mathbf{q}_2 \end{bmatrix},\ \tilde{\mathbf{v}} = \begin{bmatrix} \mathbf{v}_1 \\ \mathbf{v}_2 \end{bmatrix},\ \tilde{\mathbf{T}} = \begin{bmatrix} \mathbf{T}_1 \\ \mathbf{T}_2 \end{bmatrix},\ \tilde{\mathbf{T}}_{dis} = \begin{bmatrix} \mathbf{T}_{1dis} \\ \mathbf{T}_{2dis} \end{bmatrix}. \tag{5.27}$$

Since $\tilde{\mathbf{A}}_n \in \mathbb{R}^{4\times 4}$ is a nonsingular matrix, dynamics (5.26) can also be written as

$$\left.\begin{aligned} \dot{\tilde{\mathbf{q}}} &= \tilde{\mathbf{v}} \\ \dot{\tilde{\mathbf{v}}} &= \tilde{\mathbf{u}}_{\tilde{q}} - \tilde{\mathbf{A}}_n^{-1}\tilde{\mathbf{T}}_{dis},\ \tilde{\mathbf{u}}_{\tilde{q}} = \tilde{\mathbf{A}}_n^{-1}\tilde{\mathbf{T}}. \end{aligned}\right\} \tag{5.28}$$

The defined function vector $\mathbf{f}$ can be expressed using the Jacobian matrices of the discussed manipulators, $\mathbf{J}_1 \in \mathbb{R}^{2\times 2}$ and $\mathbf{J}_2 \in \mathbb{R}^{2\times 2}$, as

$$\mathbf{f} = \dot{\boldsymbol{\varphi}} = \begin{bmatrix} \dot{\boldsymbol{\varphi}}_s \\ \dot{\boldsymbol{\varphi}}_{rt} \end{bmatrix} = \underbrace{\begin{bmatrix} \mathbf{J}_1 & -\mathbf{J}_2 \\ \mathbf{J}_1 & \mathbf{J}_2 \end{bmatrix}}_{\mathbf{J}_f} \begin{bmatrix} \mathbf{v}_1 \\ \mathbf{v}_2 \end{bmatrix} = \mathbf{J}_f\tilde{\mathbf{v}}. \tag{5.29}$$

The matrix $\mathbf{J}_f \in \mathbb{R}^{4\times 4}$ will be denoted as the function Jacobian matrix.

The tracking error vector $\mathbf{e} \in \mathbb{R}^{4 \times 1}$ can be expressed as

$$\mathbf{e} = \begin{bmatrix} e_{xd} \\ e_{yd} \\ e_{xc} \\ e_{yc} \end{bmatrix} = \begin{bmatrix} x_1 - x_2 \\ y_1 - y_2 \\ x_1 + x_2 - 2x^{ref} \\ y_1 + y_2 - 2y^{ref} \end{bmatrix} = \begin{bmatrix} \boldsymbol{\varphi}_s \\ \boldsymbol{\varphi}_{rt} \end{bmatrix} - \begin{bmatrix} \boldsymbol{\varphi}_s^{ref} \\ \boldsymbol{\varphi}_{rt}^{ref} \end{bmatrix} = \boldsymbol{\varphi} - \boldsymbol{\varphi}^{ref}. \quad (5.30)$$

The generalized error $\boldsymbol{\sigma} \in \mathbb{R}^{4 \times 1}$ is selected as

$$\boldsymbol{\sigma} = \begin{bmatrix} \sigma_1 & \sigma_2 & \sigma_3 & \sigma_4 \end{bmatrix}^{\mathrm{T}} = \mathbf{f} - \mathbf{f}^{ref}. \quad (5.31)$$

The control goal will be achieved if the system motion is converging to the manifold $\boldsymbol{\sigma} = \mathbf{0}$.

The first-order dynamics of the generalized error is given as

$$\dot{\boldsymbol{\sigma}} = \dot{\mathbf{f}} - \dot{\mathbf{f}}^{ref}. \quad (5.32)$$

Using (5.29), (5.32) can be expressed as

$$\dot{\boldsymbol{\sigma}} = \mathbf{J}_f \dot{\tilde{\mathbf{v}}} + \underbrace{\dot{\mathbf{J}}_f \tilde{\mathbf{v}}}_{\boldsymbol{\Upsilon}} - \dot{\mathbf{f}}^{ref}. \quad (5.33)$$

Considering (5.28), (5.33) becomes

$$\dot{\boldsymbol{\sigma}} = \mathbf{J}_f \tilde{\mathbf{u}}_{\tilde{q}} - \mathbf{J}_f \tilde{\mathbf{A}}_n^{-1} \tilde{\mathbf{T}}_{dis} + \boldsymbol{\Upsilon} - \dot{\mathbf{f}}^{ref}. \quad (5.34)$$

Let us now introduce the control vector in the function space $\mathbf{u}_f \in \mathbb{R}^{4 \times 1}$, which is related with the control acceleration in the configuration space $\tilde{\mathbf{u}}_{\tilde{q}} \in \mathbb{R}^{4 \times 1}$ through the following equation

$$\tilde{\mathbf{u}}_{\tilde{q}} = \mathbf{J}_f^{-1} \mathbf{u}_f. \quad (5.35)$$

Therefore, it is assumed that $\mathbf{J}_f$ stays a nonsingular matrix for the entire time of operation. Now, dynamics (5.34) can be written as

$$\dot{\boldsymbol{\sigma}} = \mathbf{u}_f - \left[\mathbf{J}_f \tilde{\mathbf{A}}_n^{-1} \tilde{\mathbf{T}}_{dis} - \boldsymbol{\Upsilon} + \dot{\mathbf{f}}^{ref} \right]. \quad (5.36)$$

If the equivalent control is defined as

$$\mathbf{u}_f^{eq} = \mathbf{J}_f \tilde{\mathbf{A}}_n^{-1} \tilde{\mathbf{T}}_{dis} - \boldsymbol{\Upsilon} + \dot{\mathbf{f}}^{ref} \quad (5.37)$$

dynamics (5.36) can be written in a shorter form as

$$\dot{\boldsymbol{\sigma}} = \mathbf{u}_f - \mathbf{u}_f^{eq}. \quad (5.38)$$

Dynamics (5.38) is given in the form (3.142), for which the control design process is already described in Chapter 3. In this task, the control design method will be the one based on the equivalent control estimation.

From (5.38), assuming that $\boldsymbol{\sigma}$ is available and with the equivalent control modeled as $\dot{\mathbf{u}}_f^{eq} = \mathbf{0}$, the equivalent control can be estimated as

$$
\begin{aligned}
\dot{\mathbf{z}} &= \mathbf{L}\left(\mathbf{u}_f - \mathbf{z} + \mathbf{L}\boldsymbol{\sigma}\right) \\
\hat{\mathbf{u}}_f^{eq} &= \mathbf{z} - \mathbf{L}\boldsymbol{\sigma}
\end{aligned}
\tag{5.39}
$$

where $\mathbf{L} \in \mathbb{R}^{4\times 4}$ is a constant gain matrix given as

$$
\mathbf{L} = \operatorname{diag}\left(l_1,\ l_2,\ l_3,\ l_4\right),\ l_i > 0,\ i = 1,\ 2,\ 3,\ 4
\tag{5.40}
$$

while $\mathbf{u}_f^{eq} + \mathbf{L}\boldsymbol{\sigma} = \mathbf{z} \in \mathbb{R}^{4\times 1}$ is the intermediate variable in the equivalent control estimation. The control vector in the function space is selected to enforce exponential convergence, and it is given as

$$
\mathbf{u}_f = \hat{\mathbf{u}}_f^{eq} - \mathbf{D}\boldsymbol{\sigma}
\tag{5.41}
$$

with the constant diagonal matrix $\mathbf{D} \in \mathbb{R}^{4\times 4}$

$$
\mathbf{D} = \operatorname{diag}\left(d_1,\ d_2,\ d_3,\ d_4\right),\ d_i > 0,\ i = 1,\ 2,\ 3,\ 4.
\tag{5.42}
$$

After the control vector in the function space is selected, it is mapped back to the configuration space as follows

$$
\tilde{\mathbf{T}} = \begin{bmatrix} \mathbf{T}_1 \\ \mathbf{T}_2 \end{bmatrix} = \tilde{\mathbf{A}}_n \mathbf{J}_f^{-1} \mathbf{u}_f.
\tag{5.43}
$$

The inverse of the function Jacobian matrix can be calculated using the inverse Jacobian matrices of the two pantograph manipulators as

$$
\mathbf{J}_f^{-1} = \frac{1}{2}\begin{bmatrix} \mathbf{J}_1^{-1} & \mathbf{J}_1^{-1} \\ -\mathbf{J}_2^{-1} & \mathbf{J}_2^{-1} \end{bmatrix}.
\tag{5.44}
$$

In this way, it is possible to avoid calculation of the inverse matrix for a matrix of order four. Instead of that, inversion is calculated only for two Jacobian matrices of the pantograph manipulators whose order is equal to two. The inverse Jacobian matrices of the pantograph manipulators can be calculated using (5.15).

The matrices $\mathbf{C}, \mathbf{L}$, and $\mathbf{D}$ which are used in the control algorithm synthesis are given as

$$
\begin{aligned}
\mathbf{C} &= \operatorname{diag}\left(30, 30, 30, 30\right),\ \mathbf{L} = \operatorname{diag}\left(1200, 1200, 1200, 1200\right) \\
\mathbf{D} &= \operatorname{diag}\left(35, 35, 35, 35\right).
\end{aligned}
\tag{5.45}
$$

The nominal inertia matrices of the manipulators are

$$
\mathbf{A}_{1n} = \mathbf{A}_{2n} = \operatorname{diag}\left(0.3, 0.3\right).
\tag{5.46}
$$

Two experiments were made to validate the proposed control algorithm. In the first experiment reference trajectories for the x- and y-coordinates, $x^{ref}(t)$

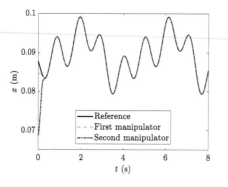

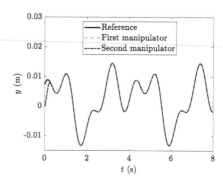

FIGURE 5.5

x-coordinate response in the first synchronization experiment.

FIGURE 5.6

y-coordinate response in the first synchronization experiment.

and $y^{ref}(t)$, were taken as combinations of sinusoidal functions, while in the second experiment, the reference trajectory in the plane was an ellipse. The sampling frequency in the experiments was 20 kHz.

The results of the first experiment are given in Figures 5.5–5.8. Both pantograph manipulators are converging to the reference trajectory and tracking is achieved (see Figures 5.5 and 5.6). All components of the tracking error vector are exponentially converging to zero, which means that the desired convergence is enforced (see Figure 5.8). The manipulators were moving along the trajectory shown in Figure 5.7. After the convergence, the tracking errors e_{xc} and e_{xd} are less than 0.22 % of the peak-to-peak amplitude of φ_{xc}^{ref}, while e_{yc} and e_{yd} are less than 0.12 % of the peak-to-peak amplitude of φ_{yc}^{ref}.

Figures 5.9–5.12 depict the results of the second experiment. As can be seen from Figures 5.9 and 5.10, both pantograph manipulators are converging to the reference trajectory and tracking is achieved. Figure 5.12 shows that all components of the tracking error vector are exponentially converging to zero, meaning that the desired convergence is enforced. Both pantograph manipulators were moving along an elliptic trajectory (see Figure 5.11). After the convergence, the tracking errors e_{xc} and e_{xd} are less than 0.17 % of the peak-to-peak amplitude of φ_{xc}^{ref}, while e_{yc} and e_{yd} are less than 0.13 % of the peak-to-peak amplitude of φ_{yc}^{ref}.

5.5   Object Manipulation Task in 2-D Space

The second task discussed in this chapter is an object manipulation task in 2-D space, for which two manipulators were again employed. Within the task, it is necessary to control the grasping force exerted on a manipulated object,

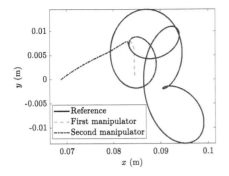

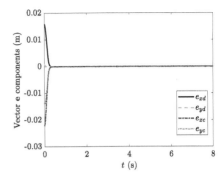

FIGURE 5.7
Manipulators' trajectories in the first synchronization experiment.

FIGURE 5.8
Vector **e** components in the first synchronization experiment.

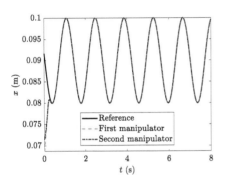

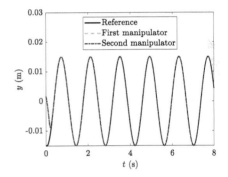

FIGURE 5.9
x-coordinate response in the second synchronization experiment.

FIGURE 5.10
y-coordinate response in the second synchronization experiment.

and position of the object in the task space. Such a task can be illustrated as in Figure 5.13, and a photo of the pantograph manipulators doing this task is shown in Figure 5.14. Each pantograph manipulator has an x-y frame attached to itself as explained in Section 5.2. Those two frames are marked with superscripts (1) and (2). The manipulated object is assumed to be grasped by the manipulators in such a way that contact points belong to two parallel faces of the object, as is illustrated in Figure 5.13.

The task that should be executed can be described as follows. One has to control the x-component of the grasping force applied to the object. The goal is to have the x-component of the grasping force at some specified value. The second goal is to have the end-effectors of the pantograph manipulators with the same y-coordinate (expressed in either of two frames) so that the y-coordinate of the grasping force does not exist and rotation around an axis

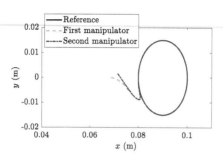

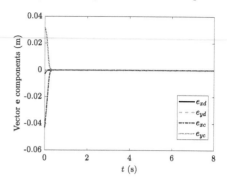

FIGURE 5.11

Manipulators' trajectories in the second synchronization experiment.

FIGURE 5.12

Vector **e** components in the second synchronization experiment.

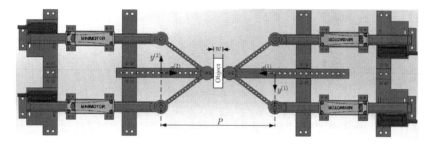

FIGURE 5.13

Two pantograph manipulators employed for object manipulation.

orthogonal to the x-y plane is prevented. In addition, the position of the object, which can be described in any of two x-y frames, has to be controlled.

Since two x-y frames exist, it is important to see how coordinates from one frame can be expressed in another frame. Here, the following notation will be used. Symbol $x_j^{(i)}$ is the x-coordinate of the jth manipulator's end-effector expressed in the ith manipulator's frame. The same notation is valid for the y-coordinate. Let us suppose that the position of the second manipulator's end-effector in the frame attached to the second pantograph manipulator is $\left[x_2^{(2)} \ y_2^{(2)} \right]^{\mathrm{T}}$. If this position is to be expressed in the frame attached to the first manipulator, then it is given as

$$\begin{bmatrix} x_2^{(1)} \\ y_2^{(1)} \end{bmatrix} = \begin{bmatrix} -x_2^{(2)} + P \\ -y_2^{(2)} \end{bmatrix} \tag{5.47}$$

where P is the distance between origins of the two frames.

Four functions are to be controlled within this task: (i) x-coordinate of the center of geometry for two end-effectors φ_{xc}, (ii) y-coordinate of the center of

FIGURE 5.14
Photo of the pantograph manipulators employed for object manipulation.

geometry for two end-effectors φ_{yc}, (iii) x-component of the grasping force φ_{gx}, and (iv) difference between y-coordinates of the manipulators φ_{yd}. For easier discussion, it is assumed that the object to be manipulated is initially placed in such a way that certain grasping force is applied to the object, meaning that contacts between the object and manipulators are established. The object dimensions and its initial placement with respect to the manipulators are assumed to be such that the grasping force exists as long as the distance between the end-effectors of the manipulators is less than w which represents the thickness of the object as illustrated in Figure 5.13. The grasping force is modeled using a spring-damper model. Four scalar functions to be controlled are expressed as follows

$$\varphi\left(\mathbf{q}_1, \mathbf{q}_2, \mathbf{v}_1, \mathbf{v}_2\right) = \begin{bmatrix} \varphi_{xc}\left(\mathbf{q}_1, \mathbf{q}_2\right) \\ \varphi_{yc}\left(\mathbf{q}_1, \mathbf{q}_2\right) \\ \varphi_{gx}\left(\mathbf{q}_1, \mathbf{q}_2, \mathbf{v}_1, \mathbf{v}_2\right) \\ \varphi_{yd}\left(\mathbf{q}_1, \mathbf{q}_2\right) \end{bmatrix} =$$

$$= \begin{bmatrix} \frac{1}{2}\left(x_1^{(1)} + x_2^{(1)}\right) \\ \frac{1}{2}\left(y_1^{(1)} + y_2^{(1)}\right) \\ K_e\left(x_1^{(1)} + x_2^{(2)} - P + w\right) + D_e\left(\dot{x}_1^{(1)} + \dot{x}_2^{(2)}\right) \\ y_1^{(1)} - y_2^{(1)} \end{bmatrix} \tag{5.48}$$

where K_e and D_e are stiffness and damping coefficients characterizing the object at the contact points between the manipulators and object. It is assumed that the pantograph manipulators will touch the object when

$x_1^{(1)} + x_2^{(2)} = P - w$. For simplicity, one can think that the grasped object is at the beginning placed in the center position between the two manipulators, and it will be grasped when they both move for $(P - w)/2$ in their positive x-direction. The functions φ_{xc}, φ_{yc}, and φ_{yd} are expressed in the frame attached to the first manipulator. Considering (5.47), the vector $\boldsymbol{\varphi}$ can also be written in the following form

$$
\boldsymbol{\varphi}\left(\mathbf{q}_1, \mathbf{q}_2, \mathbf{v}_1, \mathbf{v}_2\right) =
\begin{bmatrix}
\frac{1}{2}\left(x_1^{(1)} - x_2^{(2)} + P\right) \\
\frac{1}{2}\left(y_1^{(1)} - y_2^{(2)}\right) \\
K_e\left(x_1^{(1)} + x_2^{(2)} - P + w\right) + D_e\left(\dot{x}_1^{(1)} + \dot{x}_2^{(2)}\right) \\
y_1^{(1)} + y_2^{(2)}
\end{bmatrix}.
\tag{5.49}
$$

Assuming that $x^{ref}(t)$ and $y^{ref}(t)$ are two times differentiable functions of time, and $F_g^{ref}(t)$ is a differentiable function of time, the references for the functions are given as

$$
\boldsymbol{\varphi}^{ref}(t) =
\begin{bmatrix}
\varphi_{xc}^{ref}(t) \\
\varphi_{yc}^{ref}(t) \\
\varphi_{gx}^{ref}(t) \\
\varphi_{yd}^{ref}(t)
\end{bmatrix}
=
\begin{bmatrix}
x^{ref} \\
y^{ref} \\
F_g^{ref} \\
0
\end{bmatrix}.
\tag{5.50}
$$

The function vector will be defined, according the procedure given in Section 3.2.1, as follows

$$
\mathbf{f} =
\begin{bmatrix}
f_{xc} \\
f_{yc} \\
f_{gx} \\
f_{yd}
\end{bmatrix}
=
\begin{bmatrix}
\dot{\varphi}_{xc} \\
\dot{\varphi}_{yc} \\
\dot{\varphi}_{gx} \\
\dot{\varphi}_{yd}
\end{bmatrix}
=
\begin{bmatrix}
\frac{1}{2}\left(\dot{x}_1^{(1)} - \dot{x}_2^{(2)}\right) \\
\frac{1}{2}\left(\dot{y}_1^{(1)} - \dot{y}_2^{(2)}\right) \\
D_e\left(\ddot{x}_1^{(1)} + \ddot{x}_2^{(2)}\right) + K_e\left(x_1^{(1)} + x_2^{(2)} - P + w\right) \\
\dot{y}_1^{(1)} + \dot{y}_2^{(2)}
\end{bmatrix}.
\tag{5.51}
$$

Since $\begin{bmatrix} \dot{x}_1^{(1)} & \dot{y}_1^{(1)} \end{bmatrix}^T = \mathbf{J}_1\mathbf{v}_1$ and $\begin{bmatrix} \dot{x}_2^{(2)} & \dot{y}_2^{(2)} \end{bmatrix} = \mathbf{J}_2\mathbf{v}_2$, one can write (5.51) in the following form

$$
\mathbf{f} =
\begin{bmatrix}
\frac{1}{2} & 0 & 0 & 0 \\
0 & \frac{1}{2} & 0 & 0 \\
0 & 0 & D_e & 0 \\
0 & 0 & 0 & 1
\end{bmatrix}
\begin{bmatrix}
\mathbf{J}_1 & -\mathbf{J}_2 \\
\mathbf{J}_1 & \mathbf{J}_2
\end{bmatrix}
\begin{bmatrix}
\mathbf{v}_1 \\
\mathbf{v}_2
\end{bmatrix}
+
\begin{bmatrix}
0 \\
0 \\
K_e\left(x_1^{(1)} + x_2^{(2)} - P + w\right) \\
0
\end{bmatrix}.
\tag{5.52}
$$

According to the procedure given in Section 4.1, the reference for the function vector is

$$
\mathbf{f}^{ref} =
\begin{bmatrix}
f_{xc}^{ref} \\
f_{yc}^{ref} \\
f_{gx}^{ref} \\
f_{yd}^{ref}
\end{bmatrix}
=
\begin{bmatrix}
\dot{\varphi}_{xc}^{ref} - c_1\left(\varphi_{xc} - \varphi_{xc}^{ref}\right) \\
\dot{\varphi}_{yc}^{ref} - c_2\left(\varphi_{yc} - \varphi_{yc}^{ref}\right) \\
\varphi_{gx}^{ref} \\
\dot{\varphi}_{yd}^{ref} - c_4\left(\varphi_{yd} - \varphi_{yd}^{ref}\right)
\end{bmatrix}
\tag{5.53}
$$

where c_1, c_2, c_4 are positive constants.

Due to the grasping force applied to the object, there are reaction forces, and they introduce additional forces acting on the motors of the manipulators. These forces have to be included in the dynamic model of the manipulators, and they will appear as a part of the total disturbance. Inclusion of these forces can be done using the Jacobian matrices of the manipulators. The dynamics of the two manipulators can now be written as

$$\left.\begin{array}{c} \dot{\mathbf{q}}_1 = \mathbf{v}_1 \\ \mathbf{A}_{1n}\dot{\mathbf{v}}_1 = \mathbf{T}_1 - (\mathbf{T}_{1dis} - \mathbf{T}_1^r) \end{array}\right\} \tag{5.54}$$

$$\left.\begin{array}{c} \dot{\mathbf{q}}_2 = \mathbf{v}_2 \\ \mathbf{A}_{2n}\dot{\mathbf{v}}_2 = \mathbf{T}_2 - (\mathbf{T}_{2dis} - \mathbf{T}_2^r) \end{array}\right\} \tag{5.55}$$

where forces appearing due to the reaction forces, $\mathbf{T}_1^r = [T_{11}^r \ T_{21}^r]^{\mathrm{T}}$ and $\mathbf{T}_2^r = [T_{12}^r \ T_{22}^r]^{\mathrm{T}}$, are given by

$$\begin{array}{rcl} \mathbf{T}_1^r & = & \mathbf{J}_1^{\mathrm{T}}\begin{bmatrix} -F_{1gx} \\ -F_{1gy} \end{bmatrix} \\ \mathbf{T}_2^r & = & \mathbf{J}_2{}^{\mathrm{T}}\begin{bmatrix} -F_{2gx} \\ -F_{2gy} \end{bmatrix}. \end{array} \tag{5.56}$$

In (5.56), F_{igx} ($i = 1, 2$) is the x-component of the force applied to the grasped object by the ith manipulator, while F_{igy} ($i = 1, 2$) is the y-component of the force applied to the grasped object by the same manipulator. Now, the dynamics of the system consisting of the both manipulators is

$$\left.\begin{array}{c} \dot{\tilde{\mathbf{q}}} = \tilde{\mathbf{v}} \\ \tilde{\mathbf{A}}_n\dot{\tilde{\mathbf{v}}} = \tilde{\mathbf{T}} - \left(\tilde{\mathbf{T}}_{dis} - \tilde{\mathbf{T}}^r\right) \end{array}\right\} \tag{5.57}$$

The matrices and vectors appearing in (5.57) are given in (5.27) and by the following equation

$$\tilde{\mathbf{T}}^r = \begin{bmatrix} \mathbf{T}_1^r \\ \mathbf{T}_2^r \end{bmatrix}. \tag{5.58}$$

Using the fact that $\tilde{\mathbf{A}} \in \mathbb{R}^{4\times 4}$ is a nonsingular matrix, an alternative form of (5.57) is

$$\left.\begin{array}{c} \dot{\tilde{\mathbf{q}}} = \tilde{\mathbf{v}} \\ \dot{\tilde{\mathbf{v}}} = \tilde{\mathbf{u}}_{\tilde{q}} - \tilde{\mathbf{A}}_n^{-1}\left(\tilde{\mathbf{T}}_{dis} - \tilde{\mathbf{T}}^r\right), \quad \tilde{\mathbf{u}}_{\tilde{q}} = \tilde{\mathbf{A}}_n^{-1}\tilde{\mathbf{T}}. \end{array}\right\} \tag{5.59}$$

The tracking error vector $\mathbf{e} \in \mathbb{R}^{4\times 1}$ and the generalized error $\boldsymbol{\sigma} \in \mathbb{R}^{4\times 1}$ are defined as

$$\mathbf{e} = \begin{bmatrix} e_{xc} \\ e_{yc} \\ e_{gx} \\ e_{yd} \end{bmatrix} = \boldsymbol{\varphi} - \boldsymbol{\varphi}^{ref} \tag{5.60}$$

$$\boldsymbol{\sigma} = \begin{bmatrix} \sigma_1 \\ \sigma_2 \\ \sigma_3 \\ \sigma_4 \end{bmatrix} = \mathbf{f} - \mathbf{f}^{ref}. \tag{5.61}$$

The control goal will be achieved if the system motion is converging to the manifold $\boldsymbol{\sigma} = \mathbf{0}$.

The first-order dynamics of the generalized error is

$$\dot{\boldsymbol{\sigma}} = \dot{\mathbf{f}} - \dot{\mathbf{f}}^{ref} \tag{5.62}$$

Taking (5.52) into account, (5.62) can further be written as

$$\dot{\boldsymbol{\sigma}} = \begin{bmatrix} \frac{1}{2} & 0 & 0 & 0 \\ 0 & \frac{1}{2} & 0 & 0 \\ 0 & 0 & D_e & 0 \\ 0 & 0 & 0 & 1 \end{bmatrix} \begin{bmatrix} \mathbf{J}_1 & -\mathbf{J}_2 \\ \mathbf{J}_1 & \mathbf{J}_2 \end{bmatrix} \begin{bmatrix} \dot{\mathbf{v}}_1 \\ \dot{\mathbf{v}}_2 \end{bmatrix} + \begin{bmatrix} \frac{1}{2} & 0 & 0 & 0 \\ 0 & \frac{1}{2} & 0 & 0 \\ 0 & 0 & D_e & 0 \\ 0 & 0 & 0 & 1 \end{bmatrix} \begin{bmatrix} \dot{\mathbf{J}}_1 & -\dot{\mathbf{J}}_2 \\ \dot{\mathbf{J}}_1 & \dot{\mathbf{J}}_2 \end{bmatrix} \begin{bmatrix} \mathbf{v}_1 \\ \mathbf{v}_2 \end{bmatrix}$$
$$+ \begin{bmatrix} 0 \\ 0 \\ K_e \left(\dot{x}_1^{(1)} + \dot{x}_2^{(2)} \right) \\ 0 \end{bmatrix} - \dot{\mathbf{f}}^{ref}. \tag{5.63}$$

The dynamics (5.63) can be written in the already introduced standard form

$$\dot{\boldsymbol{\sigma}} = \mathbf{J}_f \dot{\mathbf{v}} + \boldsymbol{\Upsilon} - \dot{\mathbf{f}}^{ref} \tag{5.64}$$

where the function Jacobian matrix $\mathbf{J}_f$ and $\boldsymbol{\Upsilon}$ are

$$\mathbf{J}_f = \begin{bmatrix} \frac{1}{2} & 0 & 0 & 0 \\ 0 & \frac{1}{2} & 0 & 0 \\ 0 & 0 & D_e & 0 \\ 0 & 0 & 0 & 1 \end{bmatrix} \begin{bmatrix} \mathbf{J}_1 & -\mathbf{J}_2 \\ \mathbf{J}_1 & \mathbf{J}_2 \end{bmatrix} \tag{5.65}$$

$$\boldsymbol{\Upsilon} = \begin{bmatrix} \frac{1}{2} & 0 & 0 & 0 \\ 0 & \frac{1}{2} & 0 & 0 \\ 0 & 0 & D_e & 0 \\ 0 & 0 & 0 & 1 \end{bmatrix} \begin{bmatrix} \dot{\mathbf{J}}_1 & -\dot{\mathbf{J}}_2 \\ \dot{\mathbf{J}}_1 & \dot{\mathbf{J}}_2 \end{bmatrix} \begin{bmatrix} \mathbf{v}_1 \\ \mathbf{v}_2 \end{bmatrix} + \begin{bmatrix} 0 \\ 0 \\ K_e \left(\dot{x}_1^{(1)} + \dot{x}_2^{(2)} \right) \\ 0 \end{bmatrix}. \tag{5.66}$$

Considering (5.59), (5.64) becomes

$$\dot{\boldsymbol{\sigma}} = \mathbf{J}_f \left[\tilde{\mathbf{u}}_{\tilde{q}} - \tilde{\mathbf{A}}_n^{-1} \left(\tilde{\mathbf{T}}_{dis} - \tilde{\mathbf{T}}^r \right) \right] + \boldsymbol{\Upsilon} - \dot{\mathbf{f}}^{ref} \tag{5.67}$$

Let us introduce the control vector in the function space $\mathbf{u}_f \in \mathbb{R}^{4 \times 1}$, which is related with the control acceleration vector in the configuration space through the following equation

$$\tilde{\mathbf{u}}_{\tilde{q}} = \mathbf{J}_f^{-1} \mathbf{u}_f. \tag{5.68}$$

It is again assumed that function Jacobian $\mathbf{J}_f$ stays a nonsingular matrix for the entire time of operation. Dynamics (5.67) becomes

$$\dot{\boldsymbol{\sigma}} = \mathbf{u}_f - \mathbf{J}_f \tilde{\mathbf{A}}_n^{-1} \left(\tilde{\mathbf{T}}_{dis} - \tilde{\mathbf{T}}^r \right) + \boldsymbol{\Upsilon} - \dot{\mathbf{f}}^{ref} \tag{5.69}$$

If the equivalent control is defined as

$$\mathbf{u}_f^{eq} = \mathbf{J}_f \tilde{\mathbf{A}}_n^{-1} \left(\tilde{\mathbf{T}}_{dis} - \tilde{\mathbf{T}}^r \right) - \boldsymbol{\Upsilon} + \dot{\mathbf{f}}^{ref} \tag{5.70}$$

the first-order dynamics of the generalized error can finally be written as

$$\dot{\boldsymbol{\sigma}} = \mathbf{u}_f - \mathbf{u}_f^{eq}. \tag{5.71}$$

If (5.38) and (5.71) are compared, it can be concluded that two different tasks can be described in the same framework. Therefore, the control synthesis for this task is identical as for the previous. If $\boldsymbol{\sigma}$ is available and equivalent control is modeled as $\dot{\mathbf{u}}_f^{eq} = \mathbf{0}$ (which was also assumed in the previous task), the equivalent control can be estimated as in (5.39), and control which will enforce exponential convergence to the manifold $\boldsymbol{\sigma} = \mathbf{0}$ can be chosen as in (5.41). The matrices $\mathbf{L} \in \mathbb{R}^{4 \times 4}$ and $\mathbf{D} \in \mathbb{R}^{4 \times 4}$ have the same forms as in the previous task, and they are expressed in (5.40) and (5.42). Once the control vector $\mathbf{u}_f$ is selected in the function space, it has to be transformed back to the configuration space. This transformation is the same as for the previous task, and it is given in (5.43). The only difference is that the function Jacobian is different for two tasks. For this task, the inverse of the function Jacobian can be calculated as

$$\mathbf{J}_f^{-1} = \frac{1}{2} \begin{bmatrix} \mathbf{J}_1^{-1} & \mathbf{J}_1^{-1} \\ -\mathbf{J}_2^{-1} & \mathbf{J}_2^{-1} \end{bmatrix} \begin{bmatrix} 2 & 0 & 0 & 0 \\ 0 & 2 & 0 & 0 \\ 0 & 0 & D_e^{-1} & 0 \\ 0 & 0 & 0 & 1 \end{bmatrix}. \tag{5.72}$$

In the experiments undertaken to validate the designed control algorithm, the grasping force is not measured. Therefore, it was necessary to estimate it. One solution for that could be to use reaction force observers [45] to estimate reaction forces acting on the motors in the configuration space. Having this information, one can get estimation of the forces imposed on the object by manipulators and then obtain the grasping force. However, this approach requires good knowledge of the system dynamics, which is complicated for modeling in this case. Thus, in this study, an alternative solution is adopted. Forces $\mathbf{T}_1^r$ and $\mathbf{T}_2^r$ can be considered as parts of the overall disturbances acting on the manipulators during the grasping. Measurements showed that in our system $\mathbf{T}_1^r$ and $\mathbf{T}_2^r$ are significantly larger than $\mathbf{T}_{dis1}$ and $\mathbf{T}_{dis2}$, , i.e., $|T_{disij}| \ll |T_{ij}^r|$ $(i,j = 1,2)$. Therefore, for the estimation of the grasping force, the total disturbances acting on the manipulators are approximated only by $-\mathbf{T}_1^r$ and $-\mathbf{T}_2^r$. Thus, the manipulators are considered to be modeled as

$$\left. \begin{aligned} \dot{\mathbf{q}}_1 &= \mathbf{v}_1 \\ \mathbf{A}_{1n} \dot{\mathbf{v}}_1 &= \mathbf{T}_1 - (-\mathbf{T}_1^r) = \mathbf{T}_1 + \mathbf{T}_1^r. \end{aligned} \right\} \tag{5.73}$$

$$\left.\begin{aligned} \dot{\mathbf{q}}_2 &= \mathbf{v}_2 \\ \mathbf{A}_{2n}\dot{\mathbf{v}}_2 &= \mathbf{T}_2 - (-\mathbf{T}_2^r) = \mathbf{T}_2 + \mathbf{T}_2^r. \end{aligned}\right\} \qquad (5.74)$$

It has to be noted that simplified models (5.73) and (5.74) are adopted only for the sake of the grasping force estimation, while more correct models (5.54) and (5.55) are used for the synthesis of the control algorithm, as shown in its derivation. Since the matrices $\mathbf{A}_{1n}$ and $\mathbf{A}_{2n}$ are taken as diagonal, one can estimate both $\mathbf{T}_1^r$ and $\mathbf{T}_2^r$ using a classic disturbance observer [47, 44, 48] for each of their components. Thus, four disturbance observers are used to estimate all four components. This implementation is done using first-order low pass filters, so it will be written in s domain. Therefore, assuming that $\mathbf{v}_i$ is available and with the disturbance modeled as $\dot{\mathbf{T}}_i^r = \mathbf{0}$, $i = 1, 2$, the estimation can be written in compact form as

$$\hat{\mathbf{T}}_i^r = -(\mathbf{T}_i + g\mathbf{A}_{in}\mathbf{v}_i)\frac{g}{s+g} + g\mathbf{A}_{in}\mathbf{v}_i, \ i = 1, 2 \qquad (5.75)$$

where g is the cut-off frequency of the used filters. Then, estimation of the forces F_{igx} and F_{igy} is obtained considering (5.56) as

$$\begin{aligned} \begin{bmatrix} \hat{F}_{1gx} \\ \hat{F}_{1gy} \end{bmatrix} &= -\mathbf{J}_1^{-\mathrm{T}}\hat{\mathbf{T}}_1^r \\ \begin{bmatrix} \hat{F}_{2gx} \\ \hat{F}_{2gy} \end{bmatrix} &= -\mathbf{J}_2^{-\mathrm{T}}\hat{\mathbf{T}}_2^r. \end{aligned} \qquad (5.76)$$

The x-component of the grasping force f_{gx} that is necessary in the control algorithm is calculated as

$$f_{gx} = \frac{1}{2}\left(\hat{F}_{1gx} + \hat{F}_{2gx}\right). \qquad (5.77)$$

The distance between two frames was $P = 216.77$ mm, while the damping coefficient for the grasped object was estimated in a simple experiment to be $D_e = 38.7$ kg/s and this value is used in the control algorithm. The constants used in $\mathbf{f}^{ref}$ formulation are given by

$$c_1 = c_2 = c_4 = 20 \qquad (5.78)$$

and the matrices $\mathbf{L}$ and $\mathbf{D}$ were

$$\mathbf{L} = \mathrm{diag}\,(1200, \ 1200, \ 1200, \ 1200)\,, \ \mathbf{D} = \mathrm{diag}\,(35, \ 35, \ 35, \ 35)\,. \qquad (5.79)$$

Two experiments were made to validate the proposed control algorithm. In the first experiment, references $x^{ref}(t)$, $y^{ref}(t)$ and $F_g^{ref}(t)$ were taken as sinusoidal functions, while in the second experiment, $x^{ref}(t)$ was a combination of sinusoidal functions, while the other references were sinusoidal functions. Sampling frequency in both experiments was 20 kHz.

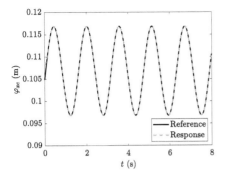

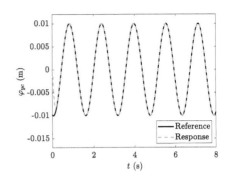

FIGURE 5.15

φ_{xc} function response in the first object manipulation experiment.

FIGURE 5.16

φ_{yc} function response in the first object manipulation experiment.

The results of the first experiment are given in Figures 5.15–5.19. The functions φ_{xc}, φ_{yc}, and φ_{gx} are converging to their references and tracking is achieved (see Figures 5.15–5.17). All components of the tracking error vector are exponentially converging to zero, which means that the desired convergence is enforced (see Figures 5.18 and 5.19). After the convergence, the error e_{xc} is less than 0.145 % of the peak-to-peak amplitude of x^{ref}, e_{yc} and e_{yd} are less than 0.3 % of the peak-to-peak amplitude of y^{ref}, while e_{gx} is less than 0.96 % of the peak-to-peak amplitude of F_g^{ref}.

For the second experiment, the results are depicted in Figures 5.20–5.24. Figures 5.20–5.22 show that the functions φ_{xc}, φ_{yc}, and φ_{gx} are converging to their references and tracking is achieved. As can be seen from Figures 5.23 and 5.24, all components of the tracking error vector are exponentially converging to zero, meaning that the desired convergence is enforced. After the convergence, the error e_{xc} is less than 0.135 % of the peak-to-peak amplitude of x^{ref}, e_{yc} and e_{yd} are less than 0.275 % of the peak-to-peak amplitude of y^{ref}, while e_{gx} is less than 0.815 % of the peak-to-peak amplitude of F_g^{ref}.

5.6 Conclusion

In this chapter, an experimental validation has been presented for the approach to motion control design for functionally related systems which is proposed in this book. The experimental setup consisted of two pantograph manipulators, that were controlled by a modular real-time control system from dSPACE. The manipulators were employed for two different tasks, a motion synchronization task and an object manipulation task. The obtained experimental results prove that both tasks can be successfully executed, which means that the proposed control design method is experimentally validated.

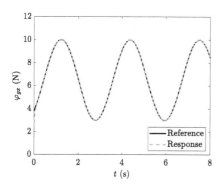

FIGURE 5.17
φ_{gx} function response in the first object manipulation experiment.

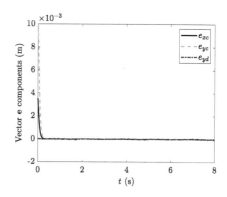

FIGURE 5.18
Vector **e** components in the first object manipulation experiment.

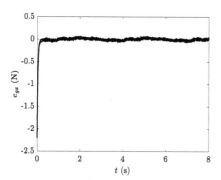

FIGURE 5.19
Tracking error for the φ_{gx} function in the first object manipulation experiment.

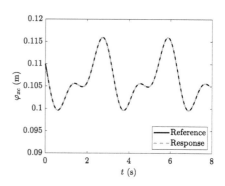

FIGURE 5.20
φ_{xc} function response in the second object manipulation experiment.

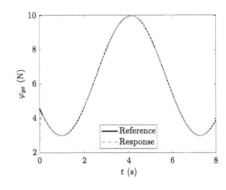

FIGURE 5.21
φ_{yc} function response in the second object manipulation experiment.

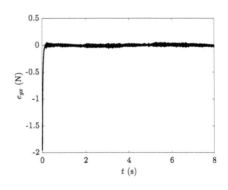

FIGURE 5.22
φ_{gx} function response in the second object manipulation experiment.

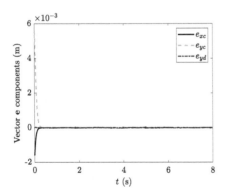

FIGURE 5.23
Vector **e** components in the second object manipulation experiment.

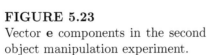

FIGURE 5.24
Tracking error for the φ_{gx} function in the second object manipulation experiment.

6

Formation Control

In this chapter, a new approach for formation control of differential-drive mobile robots is presented. The formation control in the framework of functionally related systems was presented in [77]. However, it is elaborated in more detail in this chapter. If a formation of mobile robots is analyzed, it can be treated as a group of functionally related systems. Thus, formation control of the robots can be conducted within the mentioned framework. Formation control has been the focus of many researchers in the past, due to its applications in numerous different areas. The problem of formation control for a group of mobile agents is to maintain certain relative positioning between the agents and move the created formation in a desired way. The main assumption is that a group of agents can perform a task better than a single-robot system. The advantages of a multiagent system can be recognized as reduced cost, robustness, increased efficiency, performance, and reconfigurability [63].

As mentioned above, mobile robots, subject to formation control, can be considered as a group of functionally related systems. Positions of the robots have to satisfy some relations, to enable maintaining of the formation shape and motion of the whole formation. The goal of this chapter is to discuss formation control of differential-drive mobile robots in the framework of functionally related systems. An assumption is that control system has two-layer structure. On the higher level, there is a controller calculating desired velocity vectors of the robots creating a formation, on the basis of established functional relationships between the robots. The vectors are desired velocities in a global reference frame of the plane of motion. On the lower level, each robot has a local controller which decides on translational and rotational velocity of the robot, based on the desired velocity vector and current orientation of the robot.

6.1 Low-Level Control Design

A differential-drive mobile robot is shown in Figure 6.1. The robot possesses two actuated wheels, left and right (denoted as LW and RW in the figure, respectively), and it also has the rear castor wheel (denoted as CW in the figure). The robot moves in a plane. The axes x_G and y_G define an arbitrary global reference in the plane of motion. Robot position and orientation are expressed in this frame. The robot has also attached a local reference frame, defined by the axes x_R and y_R, while the origin of this frame is the midpoint between the two actuated wheels. The midpoint coordinates x and y, and the angle θ which is measured between x_G and x_R axes, fully describe the posture of the robot. Due to its nonholonomic nature, the robot can have translational motion only in the direction of x_R axis or in the opposite direction; thus, translational velocity along the y_R axis is equal to zero. In this study, only the first-order kinematic model is treated, and it is expressed as

$$\begin{aligned} \dot{x} &= v\cos\theta \\ \dot{y} &= v\sin\theta \\ \dot{\theta} &= \omega. \end{aligned} \tag{6.1}$$

In (6.1), v is the translational velocity, while ω stands for the rotational velocity of the robot. These velocities can be trated as control inputs to the robot. In fact, one is basically controlling the rotational velocities of the right wheel and left wheel, ω_R and ω_L, respectively. Anyway, if the radius of the wheels r and distance between the wheels P are known, there exist simple relations allowing expression of v and ω in terms of ω_R and ω_L as

$$\begin{aligned} v &= \frac{r\omega_R}{2} + \frac{r\omega_L}{2} \\ \omega &= \frac{r\omega_R - r\omega_L}{P}. \end{aligned} \tag{6.2}$$

In the derivation of the control algorithm for a single robot, it is assumed that the control system has two-layer structure. A high-level controller is calculating the desired velocity vector of the robot in the inertial frame, expressed as $\mathbf{v}^{des} = \begin{bmatrix} \dot{x}^{des} & \dot{y}^{des} \end{bmatrix}^{\mathrm{T}}$, and its components are calculated based on the desired motion of the robot. An assumption is that the desired motion of the robot is described by a planar curve which is differentiable with respect to time. On the other hand, a low-level controller is determining input signals v and ω such that components of the robot velocity vector in the inertial frame, $\dot{x}$ and $\dot{y}$, satisfy the following conditions

$$\begin{aligned} \dot{x} &\xrightarrow{t\to\infty} \dot{x}^{des} \\ \dot{y} &\xrightarrow{t\to\infty} \dot{y}^{des}. \end{aligned} \tag{6.3}$$

The vector $\mathbf{v}^{des}$ can be expressed as

$$\mathbf{v}^{des} = \begin{bmatrix} v^{des}\cos\theta^{des} \\ v^{des}\sin\theta^{des} \end{bmatrix}. \tag{6.4}$$

where v^{des} and θ^{des} are the desired translational velocity and orientation of the robot. It has to be taken into account that v^{des} is not necessarily the magnitude of the vector $\mathbf{v}^{des}$, as it can be negative.

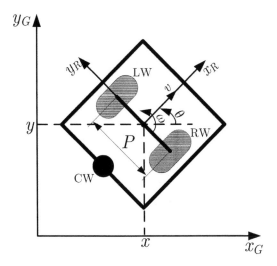

FIGURE 6.1
A differential-drive mobile robot in the global reference frame with the attached local reference frame. Modified from [77].

The desired translational velocity and orientation will be calculated as the solution of the following system of equations

$$\begin{aligned} v^{des} \cos \theta^{des} &= \dot{x}^{des} \\ v^{des} \sin \theta^{des} &= \dot{y}^{des}. \end{aligned} \tag{6.5}$$

The system of equations has always two groups of solutions, each one having an infinite number of members. If the first group solutions are expressed as $v_1^{des} = \sqrt{(\dot{x}^{des})^2 + (\dot{y}^{des})^2}$, $\theta_1^{des} = \text{atan2}\left(\dot{y}^{des}, \dot{x}^{des}\right) + 2k\pi$, $k \in \mathbb{Z}$, the solutions from the second group are $v_2^{des} = -v_1^{des}$, $\theta_2^{des} = \text{atan2}\left(\dot{y}^{des}, \dot{x}^{des}\right) + (2k+1)\pi$, $k \in \mathbb{Z}$. The angle $\text{atan2}\left(\dot{y}^{des}, \dot{x}^{des}\right)$ is in the interval $(-\pi, \pi]$. The question arises which solution to select. The translational velocity of the robot can be directly changed, as it is one control input. Yet, the orientation of the robot is being controlled by the rotational velocity ω. Thus, the most logical selection for θ^{des} is the one closest to the current orientation of the robot θ, as it requires the smallest change in the orientation. In this way, the difference between θ^{des} and θ can always stay in the range $[-\pi/2, \pi/2]$. In the case when two different solutions produce the same absolute difference between θ^{des} and θ, the one related with positive v^{des} is selected. After θ^{des} and v^{des} are chosen, it is also necessary to determine a rotational velocity profile that will enforce (6.3).

When v^{des} and θ^{des} are calculated, one can find the control signals v and ω, where the rotational velocity control input is chosen to enforce asymptotic convergence of θ to θ^{des}. Thus, the control inputs for the controlled mobile robot are given by

$$
\begin{aligned}
v &= v^{des}\\
\omega &= \dot{\theta}^{des} - K_\omega \left(\theta - \theta^{des}\right), \; K_\omega > 0.
\end{aligned}
\tag{6.6}
$$

It has to be mentioned that the presented control algorithm has been designed assuming that one wants to control the position of the robot in the plane. Due to the robot's nonholonomic nature, the orientation of the robot cannot be independently controlled and the robot will be oriented in the direction of the tangent to the robot trajectory.

The low-level controller for a differential-drive mobile robot presented in this section needs input signals $\dot{x}^{des}$ and $\dot{y}^{des}$, coming from a high-level controller, and also the orientation of the robot. The desired velocity components in the inertial frame, $\dot{x}^{des}$ and $\dot{y}^{des}$, are dependent on a particular desired motion of the robot. For example, if the robot has to track a differentiable trajectory given as a vector-valued function of time by $\mathbf{x}^{ref} = \left[x^{ref} \; y^{ref}\right]^{\mathrm{T}}$, the components can be calculated as

$$
\begin{aligned}
\dot{x}^{des} &= \dot{x}^{ref} - K_x \left(x - x^{ref}\right), \; K_x > 0\\
\dot{y}^{des} &= \dot{y}^{ref} - K_y \left(y - y^{ref}\right), \; K_y > 0.
\end{aligned}
\tag{6.7}
$$

6.2 High-Level Control Design

Mobile robots in a formation can be treated as functionally related systems. In formation control, certain functional relations between the coordinates of the robots have to be preserved in order to create a particular formation, and the whole formation has to track a reference trajectory. Assume that the formation consists of s differential-drive mobile robots, and the position of each robot in the global inertial reference frame is given by the vector $\mathbf{x}_i = [x_i \; y_i]^{\mathrm{T}}$, $i = 1, 2, \ldots, s$. The positions of all robots are described by the position vector of the formation $\mathbf{x} = [\mathbf{x}_1 \; \mathbf{x}_2 \; \ldots \; \mathbf{x}_s]^{\mathrm{T}} \in \mathbb{R}^{2s \times 1}$.

It is assumed that m functional relations have to be maintained in the system. Each relation actually represents a function, and the goal of the task is to have these functions executed, i.e., controlled to track their corresponding references. In the definition of the functions, it is assumed that definitions of certain functions can include the position of a 'virtual' mobile robot, which can for example represent the center of the formation, or it can be considered as the formation leader. The situation in which the 'virtual' mobile robot is not used is also possible. On the contrary, several 'virtual' mobile robots may exist in the formation. In this work, only the case with one mobile robot is discussed. The position of the 'virtual' robot is given by the vector $\mathbf{x}_v = [x_v \; y_v]^{\mathrm{T}}$, and it is in the same form as for the controlled differential-drive mobile robots. For the discussed case of formation control, the k-th function is expressed as $\varphi_k(\mathbf{x}, \mathbf{x}_v)$, $k = 1, 2, \ldots, m$ and it is assumed to be differentiable with respect to time. All functions can be combined in a single vector as $\varphi(\mathbf{x}, \mathbf{x}_v) = [\varphi_1(\mathbf{x}, \mathbf{x}_v) \; \varphi_2(\mathbf{x}, \mathbf{x}_v) \; \ldots \; \varphi_m(\mathbf{x}, \mathbf{x}_v)]^{\mathrm{T}}$. In the derivation of the high-level control algorithm, it will be assumed that the control signal is the vector $\dot{\mathbf{x}}$. Therefore, the relative degree of these functions with $\dot{\mathbf{x}}$ as control input is equal to one, meaning that the function vector is formed as $\varphi = \mathbf{f} \in \mathbb{R}^{m \times 1}$, i.e., it is identical to φ. The dimension of the function vector has to satisfy $m \leq 2s$, since for $m > 2s$ it is not possible to control all functions in the same time. The references for the functions are $\varphi_k^{ref}(t) = f_k^{ref}(t)$, $k = 1, 2, \ldots, m$, and they are assumed to be differentiable functions of time. Thus, the reference for the function vector is expressed as $\left[f_1^{ref}(t) \; f_2^{ref}(t) \; \ldots \; f_m^{ref}(t) \right]^{\mathrm{T}} = \mathbf{f}^{ref}(t) \in \mathbb{R}^{m \times 1}$.

The tracking error vector $\mathbf{e} \in \mathbb{R}^{m \times 1}$ and generalized error $\boldsymbol{\sigma} \in \mathbb{R}^{m \times 1}$ are identical in this case and given by

$$\mathbf{e} = \varphi - \varphi^{ref} = \boldsymbol{\sigma} = \mathbf{f} - \mathbf{f}^{ref} \tag{6.8}$$

where dependance on the position vector of the formation, position of the 'virtual' robot, and time are left out for shorter writing. The control goal is to enforce

$$\boldsymbol{\sigma} \xrightarrow{t \to \infty} \mathbf{0}. \tag{6.9}$$

The first-order dynamics of the generalized error can be written as

$$\dot{\boldsymbol{\sigma}} = \mathbf{J}_f \dot{\mathbf{x}} + \mathbf{J}_v \dot{\mathbf{x}}_v - \dot{\mathbf{f}}^{ref}, \quad \frac{\partial \mathbf{f}}{\partial \mathbf{x}} = \mathbf{J}_f \in \mathbb{R}^{m \times 2s}, \quad \frac{\partial \mathbf{f}}{\partial \mathbf{x}_v} = \mathbf{J}_v \in \mathbb{R}^{m \times 2}. \tag{6.10}$$

In (6.10), it is assumed that the function Jacobian matrix $\mathbf{J}_f$ is a full row rank matrix, i.e., $\mathrm{rank}(\mathbf{J}_f) = m$. As stated above, the vector $\dot{\mathbf{x}}$ is considered as the control signal for the generalized error dynamics (6.10), and it is determined based on the desired dynamics of the generalized error. The determined value of the vector $\dot{\mathbf{x}}$ contains the desired values of the velocity vectors for all differential-drive mobile robots, and it is denoted as $\mathbf{u}_q \in \mathbb{R}^{2s \times 1}$. These vectors are used by s low-level controllers, one for each robot, and the controllers are executing the algorithm presented in the previous section. When $\mathbf{J}_f$ has full row rank, then any desired dynamics of the generalized error can be achieved with $\mathbf{u}_q$ as the control signal. For $\dot{\mathbf{x}} = \mathbf{u}_q$ dynamics (6.10) becomes

$$\dot{\boldsymbol{\sigma}} = \mathbf{J}_f \mathbf{u}_q + \mathbf{J}_v \dot{\mathbf{x}}_v - \dot{\mathbf{f}}^{ref}. \tag{6.11}$$

Let us now introduce the control vector in the function space $\mathbf{u}_f \in \mathbb{R}^{m \times 1}$, related to the vector $\mathbf{u}_q$ by

$$\mathbf{u}_q = \mathbf{J}_f^{\#} \mathbf{u}_f + \boldsymbol{\Gamma} \dot{\mathbf{x}}_0, \quad \boldsymbol{\Gamma} = \left(\mathbf{I} - \mathbf{J}_f^{\#} \mathbf{J}_f \right) \tag{6.12}$$

where $\mathbf{J}_f^{\#} \in \mathbb{R}^{2s \times m}$ represents a right pseudoinverse of $\mathbf{J}_f$, and $\dot{\mathbf{x}}_0$ is an arbitrary velocity vector in $\mathbb{R}^{2s \times 1}$. The term $\boldsymbol{\Gamma} \dot{\mathbf{x}}_0$ is added for the case $m < 2s$, since in that case control of the functions does not require all $2s$ components of the control signal $\mathbf{u}_q$. In the case when $m = 2s$, the right pseudoinverse becomes the inverse matrix of $\mathbf{J}_f$ and then it is valid $\boldsymbol{\Gamma}\dot{\mathbf{x}}_0 = \mathbf{0}$. If the equivalent control $\mathbf{u}_f^{eq} \in \mathbb{R}^{m \times 1}$ is defined as

$$\mathbf{u}_f^{eq} = -\mathbf{J}_v \dot{\mathbf{x}}_v + \dot{\mathbf{f}}^{ref} \tag{6.13}$$

then, taking into account (6.12) and (6.13), dynamics (6.11) becomes

$$\dot{\boldsymbol{\sigma}} = \mathbf{u}_f - \mathbf{u}_f^{eq}. \tag{6.14}$$

It is important to notice that the first-order dynamics of the generalized error $\boldsymbol{\sigma}$ does not directly depend on the vector $\dot{\mathbf{x}}_0$, and it is not directly controlling the functions. The vector $\dot{\mathbf{x}}_0$ can be used for accomplishing some other goals.

If the position vectors of all the robots, including the differential-drive mobile robots and 'virtual' robot, velocity vector of the 'virtual' robot, and derivative of the function vector reference are available, then the equivalent control can be calculated. This availability is a reasonable assumption. If the assumption does not hold, the equivalent control can be estimated assuming that $\mathbf{u}_f$ and $\boldsymbol{\sigma}$ are available. With the equivalent control model $\dot{\mathbf{u}}_f^{eq} = \mathbf{0}$, the equivalent control estimation is given as

$$\begin{aligned}
\dot{\mathbf{z}} &= \mathbf{L}\left(\mathbf{u}_f - \mathbf{z} + \mathbf{L}\boldsymbol{\sigma}\right) \\
\hat{\mathbf{u}}_f^{eq} &= \mathbf{z} - \mathbf{L}\boldsymbol{\sigma}.
\end{aligned} \tag{6.15}$$

In (6.15), matrix $\mathbf{L} \in \mathbb{R}^{m \times m}$ is a constant diagonal gain matrix with positive diagonal entries, and $\mathbf{z} \in \mathbb{R}^{m \times 1}$ is the intermediate variable used in the equivalent control estimation. When the equivalent control is available, the control vector $\mathbf{u}_f$ is calculated as

$$\mathbf{u}_f = \mathbf{u}_f^{eq} - \mathbf{D}\boldsymbol{\sigma} \tag{6.16}$$

where $\mathbf{D} \in \mathbb{R}^{m \times m}$ is a constant diagonal matrix with positive diagonal entries. With this control signal, the closed-loop dynamics of the generalized error is described by

$$\dot{\boldsymbol{\sigma}} + \mathbf{D}\boldsymbol{\sigma} = \mathbf{0}. \tag{6.17}$$

From (6.17), it follows that control goal (6.9) is satisfied. Once $\mathbf{u}_f$ is calculated, the vector $\mathbf{u}_q$ is obtained as in (6.12).

It is also necessary to discuss how to select a right pseudoinverse $\mathbf{J}_f^{\#}$ in (6.12). The pseudoinverse can be selected to minimize the function $h(\mathbf{u}_q)$ given as

$$h(\mathbf{u}_q) = 0.5\mathbf{u}_q^{\mathrm{T}}\mathbf{W}\mathbf{u}_q \tag{6.18}$$

where $\mathbf{W} \in \mathbb{R}^{2s \times 2s}$ is a positive definite weighting matrix, while the minimization is done under the constraint

$$\mathbf{u}_f = \mathbf{J}_f\mathbf{u}_q. \tag{6.19}$$

In that case, the right pseudoinverse is calculated as

$$\mathbf{J}_f^{\#} = \mathbf{W}^{-1}\mathbf{J}_f^{\mathrm{T}}\left(\mathbf{J}_f\mathbf{W}^{-1}\mathbf{J}_f^{\mathrm{T}}\right)^{-1}. \tag{6.20}$$

The simplest solution is obtained when $\mathbf{W}$ is a unit matrix, and the function to be minimized in that case is the sum of squares of the vector $\mathbf{u}_q$ components, which represent desired velocities of the robots within the formation.

6.3 Simulation Results

6.3.1 Pentagon Formation

In this section, the control design procedure proposed in this chapter will be illustrated in an example. Five differential-drive mobile robots are controlled to move in a pentagon formation, with a specific placement in the vertices of the pentagon. The position of the formation, dimensions of the pentagon and its orientation have to be controlled. Different sets of functions can be selected to achieve this goal.

In this work, the following approach is used. A 'virtual' mobile robot is positioned in the center of the pentagon's circumscribed circle. The position of this robot will be defining the position of the pentagon in the global reference frame. The dimensions of the pentagon are defined by the radius of the circumscribed circle, while the angle between the x_G axis and vector $[x_1 - x_v \ y_1 - y_v]^{\mathrm{T}}$ is determining the orientation of the pentagon. The desired formation is illustrated in Figure 6.2. All notations used in the previous sections are still valid. Five differential-drive mobile robots (from MR 1 to MR 5) are positioned in the vertices of the pentagon, while the 'virtual' mobile robot (VMR) is placed in the centre of the circumscribed circle. The radius of the circle is r^{ref}, while the angle between the x_G axis and vector $[x_1 - x_v \ y_1 - y_v]^{\mathrm{T}}$ is equal to α^{ref}. In general, r^{ref} and α^{ref} are functions of time, and they are utilized for the control of the dimensions and orientation of the pentagon.

For the discussed system, the following functions are incorporated in the set of functions:

a) five functions are taken as the square of the distance between the 'virtual' mobile robot and each differential-drive mobile robot (the square of the distance is utilized for avoiding the square root function),

b) the sixth function is the oriented angle between the x_G axis and $[x_1 - x_v \ y_1 - y_v]^{\mathrm{T}}$ vector,

c) the remaining four functions are acquired by defining the oriented angle between the vector $[x_i - x_v \ y_i - y_v]^{\mathrm{T}}$ and vector $[x_{i+1} - x_v \ y_{i+1} - y_v]^{\mathrm{T}}$ for $i = 1, 2, 3, 4$.

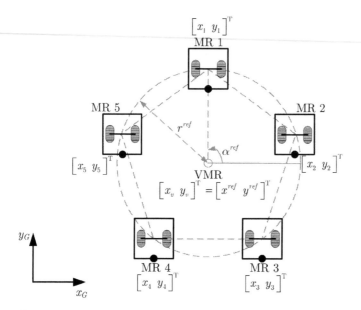

FIGURE 6.2
The desired pentagon formation. Modified from [77].

The functions are expressed as

$$\mathbf{f}\left(\mathbf{x}, \mathbf{x}_v\right) = \begin{bmatrix} f_1\left(\mathbf{x}_1, \mathbf{x}_v\right) \\ f_2\left(\mathbf{x}_2, \mathbf{x}_v\right) \\ f_3\left(\mathbf{x}_3, \mathbf{x}_v\right) \\ f_4\left(\mathbf{x}_4, \mathbf{x}_v\right) \\ f_5\left(\mathbf{x}_5, \mathbf{x}_v\right) \\ f_6\left(\mathbf{x}_1, \mathbf{x}_v\right) \\ f_7\left(\mathbf{x}_1, \mathbf{x}_2, \mathbf{x}_v\right) \\ f_8\left(\mathbf{x}_2, \mathbf{x}_3, \mathbf{x}_v\right) \\ f_9\left(\mathbf{x}_3, \mathbf{x}_4, \mathbf{x}_v\right) \\ f_{10}\left(\mathbf{x}_4, \mathbf{x}_5, \mathbf{x}_v\right) \end{bmatrix} =$$

$$= \begin{bmatrix} \left(x_1 - x_v\right)^2 + \left(y_1 - y_v\right)^2 \\ \left(x_2 - x_v\right)^2 + \left(y_2 - y_v\right)^2 \\ \left(x_3 - x_v\right)^2 + \left(y_3 - y_v\right)^2 \\ \left(x_4 - x_v\right)^2 + \left(y_4 - y_v\right)^2 \\ \left(x_5 - x_v\right)^2 + \left(y_5 - y_v\right)^2 \\ \operatorname{atan2}\left(y_1 - y_v, x_1 - x_v\right) \\ \operatorname{atan2}\left(y_1 - y_v, x_1 - x_v\right) - \operatorname{atan2}\left(y_2 - y_v, x_2 - x_v\right) \\ \operatorname{atan2}\left(y_2 - y_v, x_2 - x_v\right) - \operatorname{atan2}\left(y_3 - y_v, x_3 - x_v\right) \\ \operatorname{atan2}\left(y_3 - y_v, x_3 - x_v\right) - \operatorname{atan2}\left(y_4 - y_v, x_4 - x_v\right) \\ \operatorname{atan2}\left(y_4 - y_v, x_4 - x_v\right) - \operatorname{atan2}\left(y_5 - y_v, x_5 - x_v\right) \end{bmatrix}. \tag{6.21}$$

The vector **f** has to be differentiable with respect to time. Therefore, it is assumed, for each appearance of the atan2 function, that the positions of the robots are such that always at least one argument of the atan2 function is different from zero. From the given description of the task that has to be performed, the references for the functions are

$$
\mathbf{f}^{ref}(t) =
\begin{bmatrix}
f_1^{ref}(t) \\
f_2^{ref}(t) \\
f_3^{ref}(t) \\
f_4^{ref}(t) \\
f_5^{ref}(t) \\
f_6^{ref}(t) \\
f_7^{ref}(t) \\
f_8^{ref}(t) \\
f_9^{ref}(t) \\
f_{10}^{ref}(t)
\end{bmatrix}
=
\begin{bmatrix}
\left(r^{ref}\right)^2 \\
\left(r^{ref}\right)^2 \\
\left(r^{ref}\right)^2 \\
\left(r^{ref}\right)^2 \\
\left(r^{ref}\right)^2 \\
\alpha^{ref} \\
2\pi/5 \\
2\pi/5 \\
2\pi/5 \\
2\pi/5
\end{bmatrix}.
\tag{6.22}
$$

Each of the last five functions describes the oriented angle between two vectors in the global reference frame (the axis x_G also represents a vector). Thus, the difference $f_i - f_i^{ref}$, $i = 6, 7, \ldots, 10$ is the difference between an angle and its reference. It can be interpreted as the angle of rotation of one vector needed to make the vectors come in a desired relative position defined by f_i^{ref}. Anyway, the desired relative position can always be achieved by rotating one of the vectors for an angle in the range $(-\pi, \pi]$, and this angle can be obtained using the atan2 function as $\sigma_i = \text{atan2}\left(\sin\left(f_i - f_i^{ref}\right), \cos\left(f_i - f_i^{ref}\right)\right)$, $i = 6, 7, \ldots, 10$. However, due to the atan2 function, the generalized error components (which are equal to the tracking errors) for the five analyzed functions are discontinuous around $\pm\pi$, and they can have jumps of $\pm 2\pi$. The discontinuities can make the control of the formation unfeasible. In order to avoid that, one can keep the track about the previous value for each of the components, acquired by measurement or in simulation, and then avoid the sudden changes by subtracting or adding 2π. Thus, the generalized error components for the last five functions are determined according to the following pseudocode at each sampling interval.

1: **procedure** GENERALIZEDERRORCOMPONENTSCALCULATION
2:     $\sigma_i^{prev} \leftarrow \sigma_i$
3:     $\sigma_i \leftarrow \text{atan2}\left(\sin\left(f_i - f_i^{ref}\right), \cos\left(f_i - f_i^{ref}\right)\right)$ ▷ Valid for
       $i = 6, 7, \ldots, 10$
4:     **if** $\sigma_i - \sigma_i^{prev} > 2\pi - \epsilon$ **then**
5:         $\sigma_i \leftarrow \sigma_i - 2\pi$
6:     **else if** $\sigma_i - \sigma_i^{prev} < -2\pi + \epsilon$ **then**
7:         $\sigma_i \leftarrow \sigma_i + 2\pi$
8:     **end if**
9: **end procedure**

In the given pseudocode, σ_i^{prev} is the value of the generalized error component for the ith function from the previous sampling interval, while ϵ represents a small positive constant which can be selected such that discontinuities are detected. For the tracking errors expressed as in the given pseudocode, the time derivative is

$$\dot{\sigma}_i = \dot{f}_i - \dot{f}_i^{ref}, \ i = 6, 7, \ldots, 10. \tag{6.23}$$

Therefore, the approach for the control design stays the same as the one presented in the previous section.

Three simulations were made in MATLAB®/Simulink® to evaluate the proposed method for formation control. In the first simulation, all ten introduced functions were controlled, while the other two simulations were performed for a reduced set of functions.

For the first simulation, the control parameters were taken as $K_\omega = 30$ and $\mathbf{D} = 35\mathbf{I}^{10\times10}$. The position of the 'virtual' robot is defined by $x_v = -t$ m and $y_v = 2 + t$ m. The reference radius of the pentagon's circumscribed circle and angle describing orientation of the formation are given as $r^{ref} = \sqrt{1.5}$ m and $\alpha^{ref} = \pi/3$. The equivalent control is calculated according to (6.13). To have a realistic simulation, the components of the vector $\mathbf{u}_f$ were bounded between -150 and 150. Moreover, the input translational velocity for each robot was bounded so that its absolute value cannot exceed 30 m/s, and the input rotational velocity is bounded between -100 and 100 rad/s. The initial positions and orientations of the differential-drive mobile robots are taken as: for the first robot $(0.5$ m, 0 m, $\pi/2$ rad$)$, second robot $(5$ m, 0 m, $\pi/2$ rad$)$, third robot $(1$ m, -2 m, $\pi/2$ rad$)$, fourth robot $(-0.25$ m, -2 m, $\pi/2$ rad$)$, and fifth robot $(-1$ m, 0 m, $\pi/2$ rad$)$. Simulation results are provided in Figures 6.3–6.5. Figure 6.3 shows the motion trajectories of the controlled mobile robots and motion trajectory of the 'virtual' mobile robot. The drawn pentagons in this figure represent desired position and orientation of the formation in selected instances of time. Thus, the robots are controlled to be in the vertices of these pentagons during the entire operation. As can be seen from the shown motion trajectories, the goal is achieved. This is further confirmed if one analyzes the responses of the tracking error vector components (see Figures 6.4 and 6.5). All components are converging to zero, which demonstrates the effectiveness of the presented algorithm.

The second simulation was made to show how the controlled system would behave if only the first five functions are included in the function vector. In other words, in that scenario one wants only to control mobile robots to be positioned on the circumscribed circle of the pentagon. The right pseudoinverse of $\mathbf{J}_f$ is calculated as in (6.20), with $\mathbf{W}$ being the identity matrix. It is taken $\dot{\mathbf{x}}_0 = \mathbf{0}$ for calculation (6.12). When compared to the previous case, the matrix $\mathbf{D}$ changed its order, and now it is a square matrix of order 5 given by $\mathbf{D} = 35\mathbf{I}^{5\times5}$, while the order was 10 in the previously discussed case. All other simulation parameters had the same value as for the previous simulation. Simulation results for this case are given in Figures 6.6 and 6.7. The drawn circles in this figure represent the circumscribed circle of the pentagon.

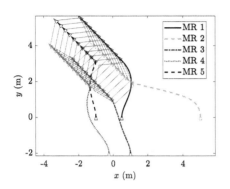

FIGURE 6.3
Motion trajectories of the mobile robots in the pentagon formation control.

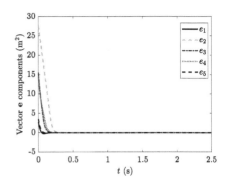

FIGURE 6.4
The first five components of the tracking error vector in the pentagon formation control.

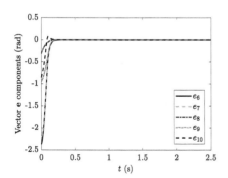

FIGURE 6.5
The last five components of the tracking error vector in the pentagon formation control.

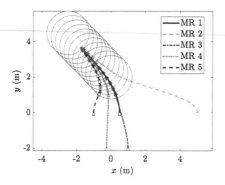

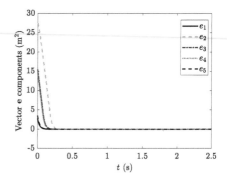

FIGURE 6.6
Motion trajectories of the mobile robots in the pentagon formation control for reduced total number of functions.

FIGURE 6.7
Components of the tracking error vector in the pentagon formation control for reduced total number of functions.

As can be seen from the shown motion trajectories, the control goal is fulfilled. This is further confirmed if the responses of the tracking error vector components are analyzed in Figure 6.7. All components are converging to zero, which demonstrates the effectiveness of the proposed control algorithm.

In addition, in order to analyze the effect of selection of the right pseudoinverse $\mathbf{J}_f^{\#}$, one more case is discussed. Still only the first five functions are included in the function vector. The right pseudoinverse matrix in (6.20) is now calculated with $\mathbf{W}$ given as

$$\mathbf{W} = \text{diag}\,(10, 1, 10, 1, 10, 1, 10, 1, 10, 1)\,. \tag{6.24}$$

This in fact means that the control system should try to give more weight to minimizing the x-components of desired velocities in the vector $\mathbf{u}_q$. Therefore, the differential-drive mobile robots should try to move more in the y-direction, while satisfying all five functions. All simulation parameters were the same as for the previous simulation. For this case, simulation results are depicted in Figures 6.8 and 6.9. If one compares trajectories from Figure 6.8 with those given in Figure 6.6, it can be concluded that the differential-drive mobile robots are moving more in the y-direction and less in the x-direction in Figure 6.8 when compared to Figure 6.6. However, the control goal is achieved, since all components of the tracking error vector are converging to zero, as shown in Figure 6.9. Therefore, the results of the last simulation demonstrate the effect introduced by the change of the matrix $\mathbf{J}_f^{\#}$. The selection of this matrix is one degree of freedom in the control design, once the number of controlled functions is less than the double value of the number of differential-drive mobile robots.

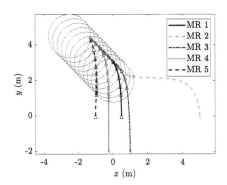

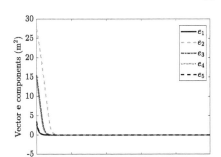

FIGURE 6.8
Motion trajectories of the mobile robots in the pentagon formation control for reduced total number of functions and changed pseudoinverse matrix.

FIGURE 6.9
Components of the tracking error vector in the pentagon formation control for reduced total number of functions and changed pseudoinverse matrix.

6.3.2 Square Formation

The performance of the proposed control method was as well tested for a square formation, depicted in Figure 6.10. For this formation, a 'virtual' mobile robot was not used. In the center of the square is the first mobile robot, determining the position of the formation. The reference trajectory for the first robot is the vector $\left[x^{ref} \ y^{ref}\right]^{\mathrm{T}}$. The dimensions of the square are defined by its side length a^{ref}, while the angle between the x_G axis and vector $\left[x_2 - x_5 \ y_2 - y_5\right]^{\mathrm{T}}$ determines the orientation of the formation and it is equal to α^{ref}. Generally, a^{ref} and α^{ref} are functions of time, and they are used for the control of the dimensions and orientation of the square formation.

For the discussed system, the following functions are defined:

a) the first two functions are $x-$ and $y-$coordinate of the first robot,

b) the four functions are taken as the square of the distance between the first robot and other robots placed in the vertices of the square,

c) the seventh function is the oriented angle between the x_G axis and $\left[x_2 - x_5 \ y_2 - y_5\right]^{\mathrm{T}}$ vector,

d) the remaining three functions are acquired by defining the oriented angle between the vector $\left[x_i - x_1 \ y_i - y_1\right]^{\mathrm{T}}$ and vector $\left[x_{i+1} - x_1 \ y_{i+1} - y_1\right]^{\mathrm{T}}$ for $i = 2, 3, 4$.

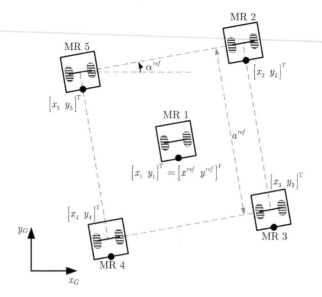

FIGURE 6.10
The desired square formation.

The functions are expressed as

$$
\mathbf{f}\left(\mathbf{x}, \mathbf{x}_v\right) =
\begin{bmatrix}
f_1\left(\mathbf{x}_1\right) \\
f_2\left(\mathbf{x}_1\right) \\
f_3\left(\mathbf{x}_1, \mathbf{x}_2\right) \\
f_4\left(\mathbf{x}_1, \mathbf{x}_3\right) \\
f_5\left(\mathbf{x}_1, \mathbf{x}_4\right) \\
f_6\left(\mathbf{x}_1, \mathbf{x}_5\right) \\
f_7\left(\mathbf{x}_2, \mathbf{x}_5\right) \\
f_8\left(\mathbf{x}_1, \mathbf{x}_2, \mathbf{x}_3\right) \\
f_9\left(\mathbf{x}_1, \mathbf{x}_3, \mathbf{x}_4\right) \\
f_{10}\left(\mathbf{x}_1, \mathbf{x}_4, \mathbf{x}_5\right)
\end{bmatrix} =
$$

$$
=
\begin{bmatrix}
x_1 \\
y_1 \\
\left(x_2 - x_1\right)^2 + \left(y_2 - y_1\right)^2 \\
\left(x_3 - x_1\right)^2 + \left(y_3 - y_1\right)^2 \\
\left(x_4 - x_1\right)^2 + \left(y_4 - y_1\right)^2 \\
\left(x_5 - x_1\right)^2 + \left(y_5 - y_1\right)^2 \\
\text{atan2}\left(y_2 - y_5, x_2 - x_5\right) \\
\text{atan2}\left(y_2 - y_1, x_2 - x_1\right) - \text{atan2}\left(y_3 - y_1, x_3 - x_1\right) \\
\text{atan2}\left(y_3 - y_1, x_3 - x_1\right) - \text{atan2}\left(y_4 - y_1, x_4 - x_1\right) \\
\text{atan2}\left(y_4 - y_1, x_4 - x_1\right) - \text{atan2}\left(y_5 - y_1, x_5 - x_1\right)
\end{bmatrix}.
$$

$$(6.25)$$

The vector **f** has to be differentiable with respect to time. The references for the functions are

$$
\mathbf{f}^{ref}(t) =
\begin{bmatrix}
f_1^{ref}(t) \\
f_2^{ref}(t) \\
f_3^{ref}(t) \\
f_4^{ref}(t) \\
f_5^{ref}(t) \\
f_6^{ref}(t) \\
f_7^{ref}(t) \\
f_8^{ref}(t) \\
f_9^{ref}(t) \\
f_{10}^{ref}(t)
\end{bmatrix}
=
\begin{bmatrix}
x^{ref} \\
y^{ref} \\
\left(a^{ref}\right)^2/2 \\
\left(a^{ref}\right)^2/2 \\
\left(a^{ref}\right)^2/2 \\
\left(a^{ref}\right)^2/2 \\
\alpha^{ref} \\
\pi/2 \\
\pi/2 \\
\pi/2
\end{bmatrix}.
\tag{6.26}
$$

For the reason already discussed, the generalized error components for the last four functions are determined based on the pseudocode provided in Section 6.3.1.

For the simulation, the control parameters were taken as $K_\omega = 30$ and $\mathbf{D} = 35\mathbf{I}^{10 \times 10}$. The reference position for the formation was $x^{ref} = t$ m and $y^{ref} = 1 + \sin t$ m. The equivalent control is calculated according to (6.13). The components of the vector $\mathbf{u}_f$, the input translational velocity for each robot, and the input rotational velocity were bounded with the same limits as in the previous sections. The initial positions and orientations of the differential-drive mobile robots are taken as: for the first robot $(0.5$ m, 0 m, $\pi/2$ rad$)$, second robot $(-2$ m, 1 m, $\pi/2$ rad$)$, third robot $(1$ m, -1.5 m, $\pi/2$ rad$)$, fourth robot $(-0.25$ m, -2 m, $\pi/2$ rad$)$, and fifth robot $(-1$ m, 0 m, $\pi/2$ rad$)$. Simulation results are given in Figures 6.11–6.14. Figure 6.11 shows the motion trajectories of the controlled mobile robots. The drawn squares in the figure stand for the desired position and orientation of the formation in selected instances of time. Therefore, the first robot is controlled to be in the center and remaining robots are controlled to be in the vertices of these squares during the operation. As can be seen from the depicted motion trajectories, the goal is achieved. This is further confirmed if one analyzes the responses of the tracking error vector components (see Figures 6.12–6.14). All components converge to zero, which demonstrates the effectiveness of the presented algorithm.

In order to make the robots move in the square formation with defined position, dimension, and orientation, a different set of functions could be selected as well. An alternative selection of the functions and their references could be

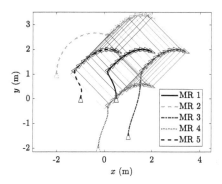

FIGURE 6.11
Motion trajectories of the mobile robots in the square formation control.

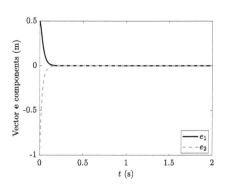

FIGURE 6.12
The first two components of the tracking error vector in the square formation control.

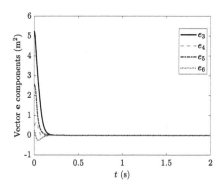

FIGURE 6.13
The third to sixth component of the tracking error vector in the square formation control.

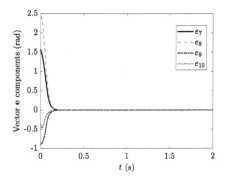

FIGURE 6.14
The last four components of the tracking error vector in the square formation control.

as follows

$$
\mathbf{f}\left(\mathbf{x}, \mathbf{x}_v\right) =
\begin{bmatrix}
f_1\left(\mathbf{x}_1\right) \\
f_2\left(\mathbf{x}_1\right) \\
f_3\left(\mathbf{x}_1, \mathbf{x}_2\right) \\
f_4\left(\mathbf{x}_1, \mathbf{x}_4\right) \\
f_5\left(\mathbf{x}_2, \mathbf{x}_3\right) \\
f_6\left(\mathbf{x}_2, \mathbf{x}_3\right) \\
f_7\left(\mathbf{x}_2, \mathbf{x}_5\right) \\
f_8\left(\mathbf{x}_2, \mathbf{x}_5\right) \\
f_9\left(\mathbf{x}_4, \mathbf{x}_5\right) \\
f_{10}\left(\mathbf{x}_4, \mathbf{x}_5\right)
\end{bmatrix}
=
\begin{bmatrix}
x_1 \\
y_1 \\
\left(x_2 - x_1\right)^2 + \left(y_2 - y_1\right)^2 \\
\left(x_4 - x_1\right)^2 + \left(y_4 - y_1\right)^2 \\
x_2 - x_3 \\
y_2 - y_3 \\
x_2 - x_5 \\
y_2 - y_5 \\
x_4 - x_5 \\
y_4 - y_5
\end{bmatrix}.
\tag{6.27}
$$

$$
\mathbf{f}^{ref}(t) =
\begin{bmatrix}
f_1^{ref}(t) \\
f_2^{ref}(t) \\
f_3^{ref}(t) \\
f_4^{ref}(t) \\
f_5^{ref}(t) \\
f_6^{ref}(t) \\
f_7^{ref}(t) \\
f_8^{ref}(t) \\
f_9^{ref}(t) \\
f_{10}^{ref}(t)
\end{bmatrix}
=
\begin{bmatrix}
x^{ref} \\
y^{ref} \\
\left(a^{ref}\right)^2/2 \\
\left(a^{ref}\right)^2/2 \\
-a^{ref}\sin\alpha^{ref} \\
a^{ref}\cos\alpha^{ref} \\
a^{ref}\cos\alpha^{ref} \\
a^{ref}\sin\alpha^{ref} \\
a^{ref}\sin\alpha^{ref} \\
-a^{ref}\cos\alpha^{ref}
\end{bmatrix}.
\tag{6.28}
$$

In this selection of functions:
  a) first two functions are the $x-$ and $y-$coordinates of the first robot,
  b) two functions are taken as the square of the distance between the first robot and second robot, as well as the first robot and fourth robot,
  c) remaining functions define relative positions between the second and third robot, the second and fifth, as well as the fourth and fifth robot within the formation.

A simulation was performed with the same input parameters as for the previous case, and simulation results are depicted in Figures 6.15–6.18. Figure 6.15 depicts the motion trajectories of the controlled mobile robots. As can be seen from the depicted motion trajectories, the control goal is again achieved. This is further confirmed if one analyzes the responses of the tracking error vector components (see Figures 6.16–6.18). All components converge to zero, which demonstrates the effectiveness of the presented algorithm for this changed set of functions as well. Since the function vector $\mathbf{f}$ is changed, the trajectories of the robots are changed as well during the convergence phase. This is, of course, the expected behavior of the robots.

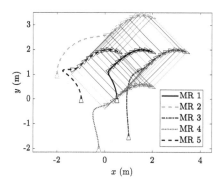

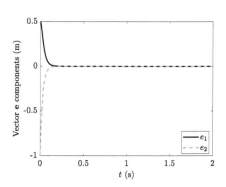

FIGURE 6.15
Motion trajectories of the mobile robots in the square formation control for different set of selected functions.

FIGURE 6.16
The first two components of the tracking error vector in the square formation control for different set of selected functions.

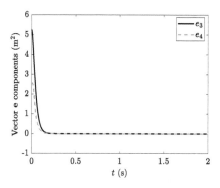

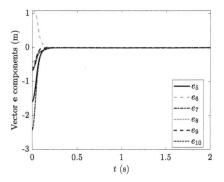

FIGURE 6.17
The second two components of the tracking error vector in the square formation control for different set of selected functions.

FIGURE 6.18
The last six components of the tracking error vector in the square formation control for different set of selected functions.

6.4 Conclusion

This chapter has presented a new approach for formation control of differential-drive mobile robots. The approach is proposed in the framework presented in this book. To make the robots move in a formation, certain functional relations, represented by functions, are established between the positions of the robots. Additionally, a 'virtual' mobile robot might be used in the definition of the functions. For example, the 'virtual' mobile robot can define the center of the formation, or it can be considered as the leader of the formation. The control algorithm is designed to enforce each of the functions to track its reference. By changing the functions, one can achieve different formation shapes and different motions of the entire formation. The proposed approach is demonstrated in simulations for a pentagon formation and a square formation of differential-drive mobile robots. It has been shown that the selection of a right pseudoinverse of the function Jacobian can be used as an additional degree of freedom in the control design for redundant tasks.

7

A New Driving Principle for Piezoelectric Walkers

This chapter presents a new approach for specifying the waveforms of driving voltages for a Piezo LEGS motor. It is motivated by the work [75] where the waveforms are synthesized using a newly proposed coordinate transformation, which was briefly redefined in the framework of functionally related systems in [78].

The approach elaborated in this chapter allows definition of the driving voltages according to the desired motion of the legs of the motor in the x- and y-directions. This concept allows a user to define the force acting on the motor rod in the y-direction and define the rod's x-direction trajectory profile. The waveforms of the driving voltages are then formed to fulfill these requirements. In addition, the presented approach enables the possibility of defining the desired step shape for the motor and, according to that definition, producing the driving voltages. Based on this idea, a simple method for Piezo LEGS motor control, identified as virtual time control, is proposed in [75], and it is used in this work as well. The method results in overshoot-free high precision positioning. The proposed approach for specifying the waveforms of driving voltages is then generalized to a piezoelectric walker with n groups of piezoelectric bimorphs.

The mentioned coordinate transformation and virtual time control in discrete form are presented briefly in [74], and utilized to implement the FPGA-based control system for the Piezo LEGS motor. In that work, only equations of coordinate transformation are given, without any discussion about constraints that have to be satisfied in order to achieve proper motor operation. The goal of that paper was to show FPGA implementation of the control algorithm, along with discussion about hardware complexity of the implemented control system. The coordinate transformation was used in [76] for force control of the Piezo LEGS motor.

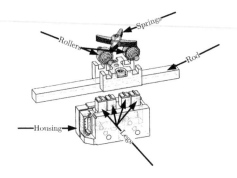

FIGURE 7.1
Mechanical construction of the Piezo LEGS motor.

7.1   Motor Description

The motor under consideration in this chapter is a commercially available Piezo LEGS motor, depicted in Figure 7.1 [50]. There exists several patents in the literature regarding the Piezo LEGS motor used in this study [30, 31, 29]. In the patents, the motor's principles of operation and driving, as well as construction method, are all discussed in detail. The motor contains four bimorph piezoelectric actuators, which are also referred to as legs; thus, the name Piezo LEGS. One leg consists of two mechanically coupled but electrically isolated groups of piezoelectric stacks, which form a multilayer d_{33} bimorph. Each group of stacks of a leg is supplied by an isolated and independent voltage. Such an electromechanical structure of piezoelectric elements allows both elongation and deflection of a leg. Hence, the planar motion of the leg can be achieved if proper voltages are supplied. The legs are driven in pairs of two, which means that in total four different voltage signals are necessary for driving the motor. Sample motion is depicted in Figure 7.2. In 1, all four legs are electrically activated. Arrows indicate the direction of the tip of each leg. In 2, the first pair of the legs is lifted and bent in the right direction, and it is keeping contact with the rod while the other pair retracts and moves in the left direction. In 3, the pair of legs that was initially retracted now extends to push against the rod while the first pair retracts. In 4, the cycle is finished as the second pair is leaning to the right while the first pair is moved left and up to come to its original position.

First, the static description of a Piezo LEGS motor leg is given and it is then followed by the dynamic model of the entire mechanical assembly. The static model contains valuable information about the relationship between the voltages applied to two halves of a leg and a static position change of that leg in the plane. However, to fully understand the operation of the Piezo LEGS motor, the development of a more intricate model is required, including

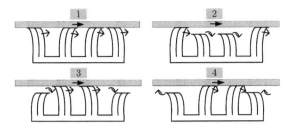

FIGURE 7.2
Illustration of the walker's motion.

dynamic behavior of the piezoelectric legs and ceramic rod. Such a model would further give insight into the nature of the interactive forces between the legs and the driving rod. Discussion on the operating principle and dynamic behavior of a Piezo LEGS motor in this chapter is short, and it contains only the information necessary for the reader to understand the proposed method for driving the motor. For a detailed dynamic modeling of the Piezo LEGS motor, the reader is advised to consider work [62]. In this work, the developed model is based on the physically meaningful parameters and macroscopically measured data for the fully assembled state of the motor. A mathematical model which describes deflection and elongation of a piezoelectric bimorph can be derived using basic principles of mechanics, namely, static equilibrium and strain compatibility between the layers of a bimorph. The model is developed based on the inverse piezoelectric effect and the Euler-Bernoulli theory of elastic deformation. A single leg is treated as a cantilever, and only the forces generated by inverse piezoelectric effect are included. This approach to modeling of a bimorph actuator is very similar to those presented in [12, 64]. Due to space constraints, intermediate steps of model development are omitted in this chapter; the final results showing the relative static deflection and elongation of the leg, driven by voltages V_1 and V_2 are

$$\Delta x = \underbrace{\frac{3}{4}\frac{d_{33}l^2}{hd}}_{k_1}(V_1 - V_2) = k_1(V_1 - V_2) \tag{7.1}$$

$$\Delta y = l\underbrace{\frac{d_{33}}{2h}}_{k_2}(V_1 + V_2) = k_2(V_1 + V_2). \tag{7.2}$$

The model is derived with reference to Figure 7.3. The constant d_{33} is a piezoelectric charge constant, l stands for length of the leg, w is its width, d represents the half width of the leg, and h is the thickness of a single piezoelectric layer. The motor rod is pressed against the motor legs with the preload springs, while a pair of roller bearings allows the movement of the rod in the direction of motion (see Figure 7.1). The driving mechanism of the motor

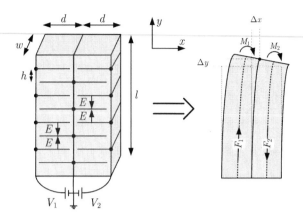

FIGURE 7.3
Leg kinematics and deflection state. © 2015 IEEE. Reprinted, with permission, from [75].

is achieved employing the phase shifted operation of pairs of legs. First, one pair of legs is brought into mechanical contact with the moving rod (grip), and then moved in the desired direction (move); after that, this pair of legs releases the rod (release) and returns to the initial position (return) (see Figure 7.2). As soon as the release operation of the first pair of legs happens, another pair of legs is brought into the grip position. If the grip-move-release-return action is denoted as a 'step', then the rod is brought into the target position by the repetition of a certain number of steps. While the above description illustrates the process, a more comprehensive understanding of driving of the Piezo LEGS motor is necessary.

Figure 7.4 shows the free body diagram of the Piezo LEGS motor. In order to maintain the figure clarity, only one pair of driving legs is depicted and indicated forces include all forces acting on the rod in that situation. It has to be noted that the motion of legs in a pair is considered to be identical. Five different forces acting on the rod are important for understanding the driving mechanism; namely, F_r is the rolling friction of the roller bearings, F_p stands for the preload force exerted on the rod by an equivalent preload spring with coefficient k_p, F_y represents the y-direction interaction force between the legs and the moving rod, F_x is the x-direction interaction force between the legs and the rod, and F_l stands for an externally applied load force in the direction opposite to the direction of the rod's motion. Depending on the magnitude of F_y and the magnitude of the force F_x, the interaction of the rod and the legs can be defined by stiction force (there is no relative motion between the legs and the rod at the point of contact) or viscous friction force (there is relative motion between the legs and the rod at the point of contact). In normal operation of the motor, there should be no relative motion between the legs and the rod and interaction is defined only by stiction force, i.e., resultant

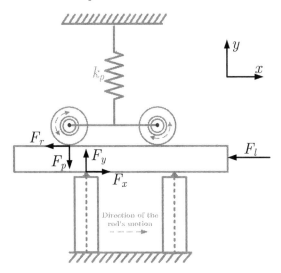

FIGURE 7.4
Free body diagram of the Piezo LEGS motor. © 2015 IEEE. Reprinted, with permission, from [75].

force applied by the legs on the rod resides inside the cone of friction. Such a situation can be accomplished by setting F_y to a high value. In that case, the deflection of the legs for Δx will cause the movement of the ceramic rod for the same amount. In this work, normal operation means no relative motion between the legs and the rod at the point of contact.

The force F_y is developed once the nonzero sum of voltages is applied to the legs. In the work [62], this force is called normal force. The rod moves in the direction of F_y when its magnitude becomes larger than the magnitude of the force F_p. The condition for the motion of the rod in the x-direction has a bit more complex nature, namely, for the motion to occur, the magnitude of F_x force needs to exceed the sum of magnitudes of the forces F_l and F_r. Anyway, in normal operation, the maximum value of force F_x is limited by the value of the stiction force threshold. The stiction force threshold is actually the product of the magnitude of F_y and stiction coefficient μ_s. Conditions for the motion of the rod in the x- and y-directions can be summarized as

$$\begin{aligned}
\text{for } \Delta y_r > 0 \quad & F_y > F_p \\
\text{for } \Delta x_r > 0 \quad & F_x > F_r + F_l \\
& \max(F_x) = \mu_s F_y.
\end{aligned} \tag{7.3}$$

In (7.3), Δx_r and Δy_r represent the change in position of the rod in x- and y-directions, respectively. The magnitude of F_y such that $\mu_s F_y = F_r + F_l$ can be denoted as the lower limit of the y-direction interaction force which creates the condition for motion; thus, it is denoted F_{lim} for purposes of further discussion.

The motion of the legs can be made such that, as soon as one pair leaves the rod, the other pair takes over and moves the rod for another full step. Different driving schemes may be proposed by changing the driving voltages supplied to the legs. The scheme proposed in this chapter does not require the exact dynamic model of the motor; nevertheless, the scheme does require (7.1) and (7.2) to hold in the Piezo LEGS operation space.

Two experiments were undertaken in order to verify the linear relation between the deflection of a leg and its driving voltages difference, as well as a relation between the elongation of a leg and its driving voltages sum, which are given by (7.1) and (7.2). The experiments are described in detail in [75] and they verified the two relations.

In the frequency range limited by achievable sampling frequency (which is approximately 20 kHz), the dynamics of the single leg is fast enough that it can be disregarded; consequently, (7.1) and (7.2) will be used to describe the dependence between the motion of the leg and its driving voltages.

7.2   Waveforms Definition based on Functions

In this section, it will be shown how driving waveforms of the motor can be designed according to the desired motion of the motor legs. It has been already noted that the motor legs are operating in pairs of two, denoted as pair 1 and pair 2. In the studies where the waveforms of the driving voltages are discussed [42, 43, 62], the authors proposed various types of the waveforms. Each type has its advantages and disadvantages concerning the motor's speed within one step and the y-direction interaction force (normal force in [62]) imposed on the rod by the legs. The motion and the forces can be analyzed once the type is selected, but no clear guidance for selection of the particular waveform has been given in these studies.

The main motivation for the investigation presented in this study is therefore the lack of clear design procedure of the supply voltages for desired motion profile of the legs. The aim is to calculate supply voltages from the motion definition coming from the desired trajectory profile and the y-direction interaction force profile. The approach presented in this work provides a very effective way to determine driving voltages that will enforce the desired step shape for a pair of the legs. Hence, the proposed approach allows us to specify two important quantities for the motor, the trajectory profile and the y-direction interaction force profile. This is possible if one can control motion of all legs in the x- and y-directions.

7.2.1 Mathematical Definition

It is assumed that (i) the movement of the first pair of legs is being controlled with voltages V_1 and V_2 , while the voltages V_3 and V_4 control motion of the second pair, (ii) two legs in a pair are controlled with the same voltages, and (iii) all legs of the motor have the same characteristics. The stated assumptions imply that displacements of legs that form one pair are identical at every instance of time.

Displacements of the first pair in x- and y-directions are given with

$$\begin{aligned} \Delta x_1\left(t\right) &= k_1\left[V_1\left(t\right) - V_2\left(t\right)\right] \\ \Delta y_1\left(t\right) &= k_2\left[V_1\left(t\right) + V_2\left(t\right)\right]. \end{aligned} \tag{7.4}$$

In (7.4), k_1 and k_2 are known constants, defined according to (7.1) and (7.2). Displacements of the legs from the second pair are

$$\begin{aligned} \Delta x_2\left(t\right) &= k_1\left[V_3\left(t\right) - V_4\left(t\right)\right] \\ \Delta y_2\left(t\right) &= k_2\left[V_3\left(t\right) + V_4\left(t\right)\right]. \end{aligned} \tag{7.5}$$

If the coefficients k_1 and k_2 were not the same for both pairs, then for the same voltage difference/sum applied to different pairs, they would have had different displacements. Such a variation in the displacement would cause a different behavior from the motor, from one step to another, depending on which pair is in control. Thus, the legs are produced with identical characteristics and it is guaranteed by the motor's manufacturer.

A problem arising from (7.4) and (7.5) is that two voltages both control x- and y-direction motion of a single pair, i.e., the motion in these two directions is coupled. From the motor's working principle, one can conclude that y-direction displacement of the legs controls y-direction interaction force (normal force) between the legs and rod. Moreover, the x-direction displacement of the legs controls the motion profile of the rod, once the normal force reaches some limit, and no relative motion exists between the legs which are driving the rod and the rod itself. Thus, it would be beneficial to have the possibility to control displacements of the legs in both directions independently. Therefore, these displacements can be considered as the functions to be controlled in order to achieve the desired driving of the motor. Let the displacements form the function vector $\mathbf{f} = \left[\Delta x_1\ \Delta y_1\ \Delta x_2\ \Delta y_2\right]^{\mathrm{T}}$, and voltages form the control vector $\mathbf{u}_q = \left[V_1\ V_2\ V_3\ V_4\right]^{\mathrm{T}}$, which is equivalent to control acceleration defined in the previous chapters. From (7.4) and (7.5), it follows

$$\mathbf{f} = \begin{bmatrix} \Delta x_1 \\ \Delta y_1 \\ \Delta x_2 \\ \Delta y_2 \end{bmatrix} = \begin{bmatrix} k_1 & -k_1 & 0 & 0 \\ k_2 & k_2 & 0 & 0 \\ 0 & 0 & k_1 & -k_1 \\ 0 & 0 & k_2 & k_2 \end{bmatrix} \begin{bmatrix} V_1 \\ V_2 \\ V_3 \\ V_4 \end{bmatrix} = \mathbf{J}_f \mathbf{u}_q. \tag{7.6}$$

The matrix $\mathbf{J}_f \in \mathbb{R}^{4\times 4}$ will be denoted as the function Jacobian. If the control vector in the function space $\mathbf{u}_f \in \mathbb{R}^{4\times 1}$ is related to $\mathbf{u}_q$ as

$$\mathbf{u}_q = \mathbf{J}_f^{-1}\mathbf{u}_f \tag{7.7}$$

then (7.6) can be written in the following form

$$\mathbf{f} = \mathbf{u}_f. \tag{7.8}$$

Considering (7.8), it is obvious that $\mathbf{u}_f$ directly controls the zeroth-order dynamics of the function vector. In previously discussed examples, one was able to directly control only the first-order dynamics of the function vector. The reason for that is the fact that functions f_i were dependent on the vector $\mathbf{v}$, and its first order dynamics can be directly set through the input force in the configuration space, and by that the first-order dynamics of $\mathbf{f}$ is enforced.

Therefore, if one wants the vector $\mathbf{f}$ to be equal to function vector reference $\mathbf{f}^{ref}$, selection of $\mathbf{u}_f$ is straightforward from (7.8), and from that it is possible to obtain $\mathbf{u}_q$ using (7.7) as

$$\mathbf{u}_q = \mathbf{J}_f^{-1}\mathbf{u}_f = \mathbf{J}_f^{-1}\mathbf{f}^{ref}. \tag{7.9}$$

The last equation can be rewritten as

$$\begin{bmatrix} V_1 \\ V_2 \\ V_3 \\ V_4 \end{bmatrix} = \frac{1}{2} \begin{bmatrix} k_1^{-1} & k_2^{-1} & 0 & 0 \\ -k_1^{-1} & k_2^{-1} & 0 & 0 \\ 0 & 0 & k_1^{-1} & k_2^{-1} \\ 0 & 0 & -k_1^{-1} & k_2^{-1} \end{bmatrix} \begin{bmatrix} \Delta x_1^{ref} \\ \Delta y_1^{ref} \\ \Delta x_2^{ref} \\ \Delta y_2^{ref} \end{bmatrix}. \tag{7.10}$$

It is obvious that by selecting the vector $\mathbf{f}^{ref}$, the desired displacements in the x- and y-directions are specified; thus, the selection of the legs trajectories in the (x, y) plane, i.e., the step shape for the motor. In this manner, it is possible to define the y-direction interaction force (normal force) profile, velocity profile within one step and step size by defining the vector $\mathbf{f}^{ref}$. The proposed approach enables definition of the legs' movement in the x-direction independently of the motion in the y-direction. The legs' x-direction trajectory profile—thus the motion profile for the rod—is determined by the vector $\mathbf{f}^{ref}$ components $\Delta x_{1/2}^{ref}$. The amplitude of these components will set step size. The motion in the y-direction and thus normal force is defined by $\Delta y_{1/2}^{ref}$ which can be set independently from $\Delta x_{1/2}^{ref}$. Hence, a decoupled selection of the trajectory in the x-direction and the normal force profile are possible.

7.2.2   Motor Constraints

In this section, motor constraints will be discussed. So-called operational constraints are defined according to the desired operation of the motor. The other group of constraints, system constraints, exists because of the PZT material characteristics and used driver properties.

Operating constraints

(O1) Both pairs of the legs have the same legs trajectory.

(O2) Trajectories are being periodically repeated.

(O3) In every time instance t, at least one pair is controlling the motion of the rod in normal operation with no slip motion between the legs and the rod.

(O4) The legs that are controlling the rod's motion are moving in the motor's desired x-direction.

System constraints

(S1) Control voltages are nonnegative.

(S2) Control voltages are smaller or equal to the supply voltage of the driver V_s.

For identical periodical trajectories of the pairs [conditions given with (O1) and (O2)], all components of the vector $\mathbf{f}^{ref}$ have to be periodic functions with period T and they have to satisfy the following relations

$$
\begin{aligned}
\Delta x_2^{ref}(t) &= \Delta x_1^{ref}(t+\tau) \\
\Delta y_2^{ref}(t) &= \Delta y_1^{ref}(t+\tau)
\end{aligned}
\tag{7.11}
$$

where τ is the time delay between the components.

As has been already discussed in Section 7.1, the condition for the normal operation motion is created if the y-direction interaction force exceeds its limit value, i.e., if the motor legs elongate for a certain length. Therefore, either pair will not be controlling the rod's motion if Δy_1^{ref} $\left(\Delta y_2^{ref}\right)$ is less than some value. If $\Delta y_1^{ref} \le k_2 c_{min}$ $\left(\Delta y_2^{ref} \le k_2 c_{min}\right)$, then the y-direction displacement of the pair 1 (pair 2) is too small to produce y-direction interaction force bigger than its limit value F_{lim}, i.e., the corresponding pair does not affect the rod's motion. The positive constant c_{min} can be experimentally determined. If elongations of both pairs are big enough to produce the y-direction interaction force bigger than F_{lim}, the pair which has a bigger y-direction elongation actually controls the rod's motion. Only in a situation when the elongations of the pairs are equal can both pairs influence the rod's motion. For each pair, the control factor cf_i $(i = 1, 2)$ can be defined as one, if the pair is controlling the rod's motion, or as zero, if that is not the case. With this definition, (O3) is satisfied if condition (7.12) is satisfied in every instance of time.

$$
\left[
\begin{array}{c}
\left(\Delta y_1^{ref}(t) > k_2 c_{min} \wedge \Delta y_1^{ref}(t) \ge \Delta y_2^{ref}(t)\right) \\
\vee \\
\left(\Delta y_2^{ref}(t) > k_2 c_{min} \wedge \Delta y_2^{ref}(t) \ge \Delta y_1^{ref}(t)\right)
\end{array}
\right]
\Leftrightarrow [cf_1 = 1 \vee cf_2 = 1].
\tag{7.12}
$$

According to the previous discussion, (7.13) is valid.

$$
\left[cf_{1/2} = 1 \Rightarrow F_{y1/2} = F_{y1/2}\left(\Delta y_{1/2}^{ref}\right)\right].
\tag{7.13}
$$

Constraint (O4) actually means that both pairs should be moving the rod in the desired direction, i.e., they must not have the opposite action in the x-direction while they are influencing the rod's motion. Vector $\mathbf{f}^{ref}$ components

Δx_1^{ref} and Δx_2^{ref} define the x-direction displacement of the legs; thus, they cause the rod to move if the pair is controlling the motion of the rod. The signs of the first derivatives of these functions define the x-coordinate moving direction of the pair. According to this fact, if the rod's motion in the positive x-direction is desired, with (O4) satisfied, (7.14) should be satisfied.

$$\left(cf_{1/2} = 1 \Rightarrow \frac{\mathrm{d}\Delta x_{1/2}^{ref}(t)}{\mathrm{d}t} > 0 \right). \tag{7.14}$$

If the time derivative in (7.14) is equal to zero, the rod is at rest, i.e., it is not moving. Similarly, if the rod's negative x-direction is desired, the condition is as follows

$$\left(cf_{1/2} = 1 \Rightarrow \frac{\mathrm{d}\Delta x_{1/2}^{ref}(t)}{\mathrm{d}t} < 0 \right). \tag{7.15}$$

System constraint (S1) will be satisfied if (7.16) is valid.

$$\left| \frac{\Delta x_{1/2}^{ref}(t)}{k_1} \right| \le \frac{\Delta y_{1/2}^{ref}(t)}{k_2}. \tag{7.16}$$

For system constraint (S2) to be satisfied, condition (7.17) has to be valid.

$$0.5 \cdot \left[\frac{\Delta y_{1/2}^{ref}(t)}{k_2} \pm \frac{\Delta x_{1/2}^{ref}(t)}{k_1} \right] \le V_s. \tag{7.17}$$

7.2.3   One Example for the Definition of Desired Motion of the Motor Legs

In this section, an example of the vector $\mathbf{f}^{ref}$ definition is given, and by that appropriate waveforms are defined as well. The definition is provided with respect to constraints and conditions that are defined in the previous section. It needs to be noted that this form of the vector $\mathbf{f}^{ref}$ is not a unique solution for its definition. Any set of $\mathbf{f}^{ref}$ components satisfying constraints from the previous section can also be utilized, which means that the number of possible solutions is unlimited. The proposed definition is simple, since all vector $\mathbf{f}^{ref}$ components are piecewise linear functions. This form was chosen in order to satisfy two conditions: 1) resultant force applied by the legs on the motor rod resides inside the cone of friction, and 2) constant rod velocity within one step. These two conditions, along with those formulated in the previous section will be satisfied if Δy_1^{ref}, Δx_1^{ref}, Δy_2^{ref}, and Δx_2^{ref} are defined as follows

$$\Delta x_1^{ref}(t) = \begin{cases} k_1 \left[-\frac{b}{2} + \frac{8}{5T} b \left(t - kT \right) \right], & kT \le t < \frac{5T}{8} + kT \\ k_1 \left[\frac{b}{2} - \frac{8}{3T} b \left(t - \frac{5T}{8} - kT \right) \right], & \frac{5T}{8} + kT \le t < T + kT \end{cases} \tag{7.18}$$

$$\Delta x_2^{ref}(t) = \Delta x_1^{ref}\left(t + \frac{T}{2}\right) \tag{7.19}$$

$$\Delta y_1^{ref}(t) = \begin{cases} k_2\left[c_{min} + \frac{8(d-c_{min})}{T}(t-kT)\right], & kT \le t < \frac{T}{8} + kT \\ k_2 d, & \frac{T}{8} + kT \le t < \frac{T}{2} + kT \\ k_2\left[d - \frac{8(d-c_{min})}{T}\left(t - \frac{T}{2} - kT\right)\right], & \frac{T}{2} + kT \le t < \frac{5T}{8} + kT \\ k_2 c_{min}, & \frac{5T}{8} + kT \le t < T + kT \end{cases} \tag{7.20}$$

$$\Delta y_2^{ref}(t) = \Delta y_1^{ref}\left(t + \frac{T}{2}\right). \tag{7.21}$$

In (7.18)–(7.21), b and d, where $d > c_{min}$, are constants which define amplitudes of the vector $\mathbf{f}^{ref}$ components, and T is their period. All components are defined as piecewise linear functions; hence, their realization is very simple. In order to satisfy condition (7.16), it has to be

$$\frac{|b|}{2} \le c_{min}. \tag{7.22}$$

To satisfy (7.17), one needs to choose d and b so that (7.23) holds.

$$\frac{d}{2} + \frac{3|b|}{20} \le V_s. \tag{7.23}$$

Resultant force applied by the legs to the motor rod will reside inside the cone of friction because (7.24) is satisfied.

$$cf_1 \cdot \Delta y_1^{ref} + cf_2 \cdot \Delta y_2^{ref} > k_2\left[c_{min} + (d - c_{min})/2\right]. \tag{7.24}$$

Besides this fact, the y-direction interaction force will remain constant for most of the period, excluding the intervals when one pair is moving down in the y-direction in order to release the rod, while another pair is moving up to take over the rod. These time intervals are now equal to one fourth of the period, but they can be made shorter with a different definition of the moving vector. The proposed definition ensures that the y-direction interaction force remains well above the limit value during the entire operation, for d enough larger than c_{min}.

The rod will be moving with a constant velocity during one step as each pair that controls the motion is moving with the same constant velocity in the x-direction. Step size will be determined by the constant b, since total x-direction deflection within one step is directly proportional to b.

The components that form vector $\mathbf{f}^{ref}$, defined in (7.18)–(7.21), are depicted in Figure 7.5 with a period of time duration.

The following fact has to be noticed. The first pair is controlling the rod's motion in the interval from $T/16$ to $9T/16$, since in this interval its elongation is larger than the elongation of the second pair, and the first pair is creating

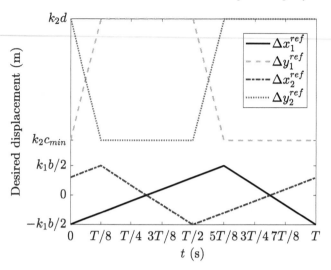

FIGURE 7.5
Desired motion of the legs.

the y-direction interaction force higher than its limit value. During this time interval, Δx_1^{ref} is rising, i.e., the first pair is moving the rod in the positive x-direction. From 0 to $T/16$ and from $9T/16$ to $5T/8$, the first pair is also moving in the positive x-direction. However, in this interval the first pair is not influencing the rod's motion. From $5T/8$ to T, Δx_1^{ref} is decreasing, the first pair is moving in the negative x-direction. If that pair would be influencing the rod's motion, it would be dragging the rod backwards. Anyhow, during that interval, the second pair is controlling the rod, and the first pair is not dragging the rod backwards. When the first pair moves to the starting position for $t = T$, a new step will start all over again. A similar discussion can be made for the second pair. If the motion in another direction is desired, one needs to change only the sign of the constant b.

7.2.4 Differences Between the Proposed Method and Existing Solutions

There are several approaches reported in the literature for driving waveforms definition, and differences between those and the approach reported here will be discussed. In the study [42], two types of driving waveforms types are mentioned, called symmetric and asymmetric sine waveforms (ASWs). Driving voltages for one leg in the first type are defined as two sine signals having the same amplitude and phase shift between them. The authors in [19] showed for this type of driving voltages that step size and normal force generation depend on amplitudes of the signals and phase shift between the signals. A fourth-order Fourier series defines asymmetric driving voltages in [42], and they are in fact periodic signals very similar to sine signals. The major difference is that during one period, the signal's positive part has longer duration than the negative one does; this enables overlapping between the two pairs. Step size and normal force are functions of the amplitudes of driving voltages and phase shift between them. The velocity profile within one step for the first type of driving waveforms will be a sine signal. For the other type, it will be a series of sine signals. It has to be noticed that for both driving voltages variants, amplitude and phase shift influence the x-direction trajectory profile and also the normal force. Due to that fact, it is not possible to independently define them so as to achieve desired normal force and x-direction trajectory profile.

In [43], the authors did waveform optimization to achieve constant stage velocity and to decrease the slip between the stage and legs. A fourth-order Fourier series describes driving signals obtained as the result of an optimization procedure to achieve constant velocity of the stage driven by a Piezo LEGS motor. The given procedure considers only stage velocity while the normal force is not discussed and it will appear as a consequence of the applied waveforms. Thus, using this approach one cannot achieve a decoupled definition of normal force and velocity profile. Moreover, a different optimization procedure has to be performed to achieve different objectives. Accordingly, a model of the motor and the driven stage or intensive measurements are required. Hence, an optimization approach is not a straightforward method for design of the waveforms.

Force waveforms are presented in [62]. They are designed to improve Piezo LEGS motor performance under high y-direction preload forces. However, velocity profile and exact normal force profile are not discussed for this definition of the waveforms. The analysis of the force waveforms definition reveals that specification of the normal force independently of the x-direction trajectory profile is not achievable.

When compared with the research cited, the approach presented in this chapter allows a waveforms definition based on the desired trajectories of the motor legs. The approach enables specification of the motion in the x-direction independently of the motion in the y-direction. Accordingly, it is possible to form driving voltages according to the desired x-direction trajectory profile and y-direction interaction force profile. This concept also gives a possibility to optimize the legs' motion so that power consumption is reduced. To be specific, when a pair is not controlling the rod's motion, its trajectory can be defined arbitrarily, so that power consumption is reduced. After that, driving waveforms which provide this behavior of the pair can be easily synthesized according to the proposed idea.

In this chapter, the proposed driving waveforms are just an example. The goal of this study is to present a straightforward method which enables a definition of driving waveforms for a Piezo LEGS motor according to the desired normal force imposed to the rod by the legs and the desired x-direction trajectory of the legs. The presented approach based on specific functions gives the possibility for the synthesis of driving waveforms that produce the desired trajectory of PZT legs, and desired motion in the x- and y-directions can be achieved. Therefore, control of the x- and y-directions interaction forces between the legs and the rod is achievable.

It is worth pointing out that proposed waveforms (PW) have limitations. Actually, they are not described as smooth signals. As the motor driver has limited bandwidth, it will not be possible to apply these waveforms in their exactly desired form to a walker. Thus, the desired requirements may not be satisfied fully. Nevertheless, the approach presented here allows for the easy definition of different waveforms so that other constraints, like the driving electronics bandwidth, can be taken into account.

To demonstrate the effectiveness of our concept, three simple experiments were made. As mentioned, the proposed driving voltages are defined to ensure the constant velocity of the Piezo LEGS motor during one step. In order to test whether that requirement is satisfied and compare with other types of waveforms, the motor was driven for three seconds by waveforms with constant parameters, and with average velocity of around 600 μm/s. The identical experiment is repeated for each of these waveforms: 1) PW, 2) sine waveforms (SW), discussed in [29, 42, 19], and 3) ASW, which are proposed in [42]. For each type, the relative error was calculated by $(\bar{v} - v)/\bar{v}$ where $\bar{v}$ denotes average velocity calculated exactly for each experiment, while v stands for velocity calculated from encoder reading and then filtered by a low pass filter. Even though the experiments lasted 3 s, parts of the responses, 0.5 s long, are given for a better view in Figure 7.6. It can be concluded that the PWs provide the smallest relative error. The mean absolute errors for the shown part of responses are 0.162, 0.196, and 0.05 for SWs, ASWs, and PWs, respectively. The response of PWs is the closest to a constant velocity response. One can explain the obtained error by the limited bandwidth of the used driver.

7.3 Control Method

This section presents the control method design based on the approach for driving waveforms definition given in the previous section. A simple controller called the virtual time controller, proposed originally in [75], is employed in this study. The pseudocode of the algorithm executed by the controller is given as follows.

```
1: procedure VIRTUALTIMECONTROL
2:     while 1 do
3:         x_r^{ref} ← ReadReference()
4:         x_r^{meas} ← ReadEncoder()
5:         e ← x_r^{ref} − x_r^{meas}
6:         |e| ← abs(e)
```

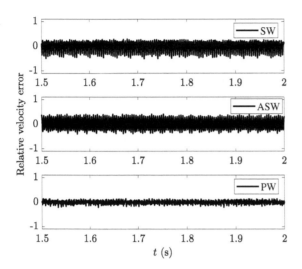

FIGURE 7.6
Relative velocity error response.

7:      $f_{old} \leftarrow f$
8:      **if** $|e| \leq e_{down}$ **then**
9:          $f \leftarrow f_{min}$
10:     **else if** $e_{down} < |e| < e_{up}$ **then**
11:             $f \leftarrow f_{min} + (f_{max} - f_{min}) \left(\frac{|e_p| - e_{down}}{e_{up} - e_{down}} \right)^2$
12:     **else**
13:         $f \leftarrow f_{max}$
14:     **end if**
15:     $T \leftarrow 1/f$
16:     **if** $e > e_{max}$ **then**
17:         $t_v \leftarrow t_v + T_s$
18:     **else if** $e < -e_{max}$ **then**
19:         $t_v \leftarrow t_v - T_s$
20:     **end if**
21:     $t_v \leftarrow t_v \cdot f_{old}/f$
22:     **if** $t_v > T$ **then**
23:         $t_v \leftarrow t_v - T$
24:     **else if** $t_v < 0$ **then**
25:         $t_v \leftarrow t_v + T$
26:     **end if**
27:   **end while**
28: **end procedure**

The control algorithm execution begins with calculation of the positioning error e according to the reference position of the rod and rod's position measured from the encoder. The next step is calculation of the positioning error's absolute value $|e|$. Then, the old value of the frequency of the vector $\mathbf{f}^{ref}$ components is saved as f_{old}, and their new frequency f is calculated in the lines 8–14. These lines perform the frequency modulation to provide a higher step frequency (meaning higher speeds and larger displacements between two sampling intervals) for larger errors, and also lower frequency for smaller errors; thus, high precision positioning is enabled. In this pseudocode, e_{down}, e_{up}, f_{min}, and f_{max} are constants that can be arbitrarily defined.

The next steps in the control algorithm actually execute the control function. They act on virtual time t_v, which is the reason for naming the controller virtual time controller. Now, the operating principle of the controller will be explained.

The virtual time is the argument of the vector $\mathbf{f}^{ref}$ components. That time is kept in the range $[0, T]$, and the algorithm basically operates on a one period long time interval. At a time instance t_v, the controller computes error, and to decrease it, it moves alongside the vector $\mathbf{f}^{ref}$ components, in the forward or backward direction, and it is done by increasing or decreasing virtual time for the sampling period T_s. If the error is positive and higher than e_{max}, the controller advances forward, moving the motor legs in the positive x-direction in order to decrease error. If the error has negative value less than $-e_{max}$, the opposite operation is executed. When the error is in the range $[-e_{max}, e_{max}]$, the virtual time is kept at a constant value, which means that the legs keep their current position without moving. In the undertaken experiments, e_{max} was taken equal to two encoder pulses, since ±1 pulse is the expected steady-state error caused by the used encoder. This basically means that the controller has a dead zone. The next step in the control algorithm is virtual time scaling. This operation is executed in order to prevent big alterations of the vector $\mathbf{f}^{ref}$, i.e., considerable alterations of the driving voltages, that can distract smooth moving. These substantial alterations can occur if the vector $\mathbf{f}^{ref}$ frequency had a huge change, when compared to its value in the previous sampling interval. Then, even a small change in the virtual time value can cause huge jumps in the values of the driving voltages. The next two steps secure that virtual time stays in the range $[0, T]$. Basically, these two steps are not even necessary, since the components of the vector $\mathbf{f}^{ref}$ are defined as periodic functions with period T. The amplitudes of these components, i.e., constants b and d, are defined before the controller starts its operation and they are kept constant during its work time.

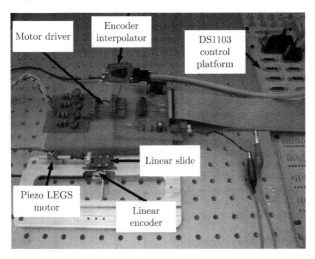

FIGURE 7.7
Experimental setup for the positioning experiments.

7.4 Experimental Results

The experimental setup, shown in Figure 7.7, consisted of the linear slide Schneeberger ND with caged rollers, connected to the motor rod. The slide includes position measurement in the mechanical structure. The position feedback is obtained by a high quality optical encoder Mercury 3500 from MicroE Systems. The encoder is used in combination with an interpolator that provides measurement resolution of 5 nm. The motor is driven using a custom-designed PWM-based switching driver [20]. Driving algorithms for both experiments are implemented in C programming language and executed on a DS1103 control platform produced by dSPACE.

Three positioning experiments were done to validate the proposed approach for driving waveforms synthesis and the presented virtual time controller. The results of those experiments are shown in this section. The used encoder has a declared measurement resolution of 5 nm, but it is important to note that its measured value oscillates ±1 pulse (i.e., ±5 nm) in a steady state, which was confirmed by the manufacturer. The parameters of the vector $\mathbf{f}^{ref}$ components and control algorithm parameters, that are all mentioned in the previous sections, are provided in Table 7.1. The virtual time controller has a dead zone of ±2 encoder pulses, that corresponds to ±10 nm. Due to the dead zone, the steady-state error of ±2 encoder pulses is expected in the experiments. The perfect controller would have a steady state error of ±1 encoder pulse because of the encoder's characteristics. Thus, the desired

TABLE 7.1
Values of experimental parameters.

Parameter	Value
c_{min}	18.5 V
b	34 V
d	50 V
V_s	46 V
e_{down}	50 encoder pulses
e_{up}	1000 encoder pulses
f_{min}	5 Hz
f_{max}	500 Hz
e_{max}	2 encoder pulses
T_s	50 μs

positioning accuracy is just 5 nm or one encoder pulse bigger than the theoretically achievable one.

The first experiment was undertaken for the reference position chosen as a step signal with the amplitude of 40 encoder pulses, which is in fact 200 nm. The obtained response is given in Figure 7.8. The presented controller showed satisfactory performance, since the motor achieved settling time of less than 0.1 s, with no overshoot, and with steady-state error within expected ± 2 encoder pulses. Part of the response between 0.6 and 0.7 s is zoomed and given separately in Figure 7.10 in order to justify achieved accuracy. Part of the response between 0 and 0.1 s (see Figure 7.9) is proving the encoder's inevitable steady state error, since it is obvious that output of the encoder oscillates even when the motor is not moving, which is the case when positioning error was within ± 2 encoder pulse, as constant control voltages were applied in those time intervals. For the reference given, the motor's displacement is less than one step, and the motor moved with a constant frequency of 5 Hz according to the given pseudocode of the control algorithm.

In the second performed experiment, the reference position was selected as a step signal with the amplitude equal to a hundred thousand encoder pulses. The response result is shown in Figure 7.11. The motor reaches a steady state in about 0.7 s, without overshoot. Residual steady state error is again within ± 2 encoder pulses, because of the previously mentioned reasons. The zoomed part of the response (see Figure 7.12) illustrates that the accuracy within 2 encoder pulses is again achieved. Frequency modulation influence can be noticed from 0.935 to 1.045 s. The controller was able to preserve the same accuracy for a 2500 times bigger position reference.

One more experiment was done, and the reference position was a square wave. The peak-to-peak amplitude and frequency of the square wave were selected as twenty thousand encoder pulses and 1 Hz, respectively. The purpose of this experiment was to examine the controller's repeatability. The response depicted in Figure 7.13 is characterized with a short settling time, no overshoot, and a small steady-state error; thus, these response characteristics

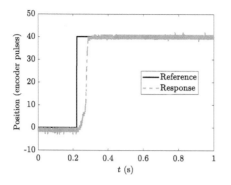

FIGURE 7.8
Position response for the first experiment.

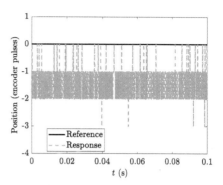

FIGURE 7.9
Position response for the first experiment - detailed view 1.

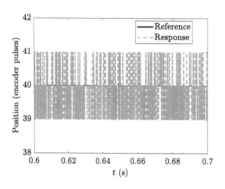

FIGURE 7.10
Position response for the first experiment - detailed view 2.

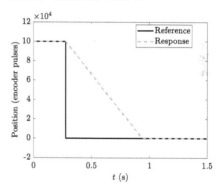

FIGURE 7.11
Position response for the second experiment.

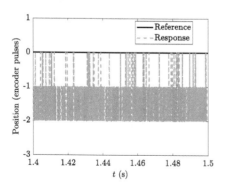

FIGURE 7.12
Position response for the second experiment - detailed view.

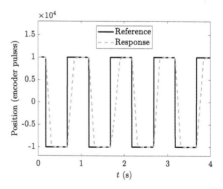

FIGURE 7.13
Position response for the third experiment.

stayed the same as in the previous two experiments. The steady-state error within ± 2 encoder pulses is again present, and its cause has been already explained.

The proposed control method is considerably simpler for implementation than the methods proposed in [19] and [42]. Obtained experimental results are showing satisfactory positioning accuracy, without overshoot and expected steady state error within ± 2 encoder pulses (± 10 nm). This is improvement in comparison with the results presented in [19] and [42]. In [19] and [42], position responses had overshoot. Results presented in [19] had much greater steady-state error when compared to those obtained with the virtual time controller, while in [42], that error was ± 5 nm, in comparison with ± 10 nm presented in this study. In this study, it was shown that error value remained the same for any reference size. Anyway, the main focus of the study was to present a new method for driving waveforms synthesis; therefore, numerous different control methods can be proposed based on this approach.

All experimental results are given without load force acting on the motor's rod in the x-direction. A load force's counteracting the driving force is actually changing the limit value of the y-direction interaction force, i.e., for higher load force the constant c_{min} must have a higher value to have the operating constraint (O3) satisfied. Therefore, the stiction force threshold has to be increased for higher load forces if normal operation is desired. This can be done by increasing the y-direction interaction force; in addition, the presented approach allows it, and it is done by changing the amplitude of $\Delta y_{1/2}^{ref}$. In the proposed definition of the vector $\mathbf{f}^{ref}$, stiction force threshold is defined by constants c_{min} and d through (7.13) and (7.24). If constant c_{min} is experimentally determined for one value of the load force, then an increase of the load force will change its value to $c_{min}^* > c_{min}$. However, to have no slip motion between the legs and the rod, finding c_{min}^* is unnecessary. It is enough to have the left side of (7.24) greater than $k_2 c_{min}^*$ in every instance of time; this can be achieved by increasing the constant d. On the other hand, if a load force is acting in the direction of the driving force, it will decrease the value of constant c_{min}, i.e., it will decrease the stiction force threshold. However, all operating and system constraints will still be satisfied for the proposed definition of the vector $\mathbf{f}^{ref}$ and no change in proper operation will occur.

7.5 Generalization of the Presented Approach

The proposed approach for driving waveforms synthesis can be generalized for a motor with n groups of legs, where each group can contain a different number of legs. Let us assume that the first group is supplied by the voltages V_1 and V_2, the second by V_3 and V_4, ith group ($i = 1, 2, \ldots, n$) by V_{2i-1} and V_{2i}, and the last group by V_{2n-1} and V_{2n}. One would like to independently control displacements of all groups in the x- and y-directions. Displacements of the ith group in x- and y-directions are given by

$$\Delta x_i (t) = k_1 [V_{2i-1} (t) - V_{2i} (t)]$$
$$\Delta y_i (t) = k_2 [V_{2i-1} (t) + V_{2i} (t)]. \qquad (7.25)$$

Let the displacements form the function vector $\mathbf{f} = [\Delta x_1 \ \Delta y_1 \ldots \Delta x_n \ \Delta y_n]^{\mathrm{T}}$, and voltages form the control vector $\mathbf{u}_q = [V_1 \ V_2 \ldots V_{2n-1} \ V_{2n}]^{\mathrm{T}}$. From (7.25), it can be written

$$\mathbf{f} = \begin{bmatrix} \Delta x_1 \\ \Delta y_1 \\ \Delta x_2 \\ \Delta y_2 \\ \vdots \\ \Delta x_n \\ \Delta y_n \end{bmatrix} = \begin{bmatrix} k_1 & -k_1 & 0 & 0 & \ldots & 0 & 0 \\ k_2 & k_2 & 0 & 0 & \ldots & 0 & 0 \\ 0 & 0 & k_1 & -k_1 & \ldots & 0 & 0 \\ 0 & 0 & k_2 & k_2 & \ldots & 0 & 0 \\ & & & \vdots & & & \\ 0 & 0 & 0 & 0 & \ldots & k_1 & -k_1 \\ 0 & 0 & 0 & 0 & \ldots & k_2 & k_2 \end{bmatrix} \begin{bmatrix} V_1 \\ V_2 \\ V_3 \\ V_4 \\ \vdots \\ V_{2n-1} \\ V_{2n} \end{bmatrix} = \mathbf{J}_f \mathbf{u}_q.$$

$$(7.26)$$

One can note that (7.6) is a special case of (7.26), written for $n = 2$ groups, each containing two legs. Therefore, the proposed approach for driving waveforms synthesis can be again employed. Thus, by selecting the vector $\mathbf{f}^{ref}$, and having $\mathbf{u}_q = \mathbf{J}_f^{-1} \mathbf{f}^{ref}$, desired displacements of each group in the x- and y-directions can be independently specified.

7.6 Conclusion

In this chapter, a novel approach that can be used for the Piezo LEGS motor waveforms definition has been presented. Further, the approach has been extended to a motor operating with n groups of piezoelectric bimorphs (legs). The presented approach gives the possibility for defining driving voltages according to the driving characteristics and desired step shape. In addition, the approach simplifies design of the control system, as shown by the presented controller that showed good control performance, by having steady-state error within ± 2 encoder pulses, when the controller was applied to a Piezo LEGS motor.

8

Illustrative Simulation Examples

In this chapter, the control synthesis approach presented in Chapter 3 will be illustrated on several illustrative simulation examples, for tasks involving robotic manipulators. Two planar manipulators will be controlled in a motion synchronization task, and also in an object manipulation task. The same approach will be applied for a bilateral system. Finally, it is also demonstrated how the same approach can be utilized in an object manipulation task in three-dimensional (3-D) space.

8.1 Tasks Involving Planar Manipulators

In this section, control for two different tasks that employ two planar manipulators will be considered. These tasks are: (i) motion synchronization in 2-D space and (ii) object manipulation in 2-D space.

8.1.1 Direct Kinematics and Dynamic Model of Planar Manipulator

A planar manipulator can be represented as in Figure 8.1. Direct kinematics of the manipulator and its dynamic model are taken from [59]. Notation is used as follows. For $i = 1, 2$, m_i denotes the mass of link i; l_i is the length of link i; l_{ci} stands for the distance from the previous joint to the center of mass of link i; and I_i is the moment of inertia of link i about an axis coming out of the page, passing through the center of mass of link i.

The position of the end-effector in the inertial x-y frame (task space) $\mathbf{x} = [x \; y]^{\mathrm{T}}$ can be expressed in terms of the configuration vector $\mathbf{q} = [q_1 \; q_2]^{\mathrm{T}}$ as

$$\mathbf{x} = \begin{bmatrix} x \\ y \end{bmatrix} = \begin{bmatrix} l_1 \cos q_1 + l_2 \cos (q_1 + q_2) \\ l_1 \sin q_1 + l_2 \sin (q_1 + q_2) \end{bmatrix}. \tag{8.1}$$

When the last equation is differentiated with respect to time, velocity of the end effector in the task space is

$$\dot{\mathbf{x}} = \begin{bmatrix} \dot{x} \\ \dot{y} \end{bmatrix} = \begin{bmatrix} -l_1 \sin q_1 - l_2 \sin (q_1 + q_2) & -l_2 \sin (q_1 + q_2) \\ l_1 \cos q_1 + l_2 \cos (q_1 + q_2) & l_2 \cos (q_1 + q_2) \end{bmatrix} \begin{bmatrix} \dot{q}_1 \\ \dot{q}_2 \end{bmatrix} = \mathbf{Jv}. \tag{8.2}$$

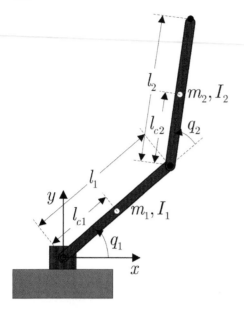

FIGURE 8.1
Planar manipulator.

The matrix $\mathbf{J}$ in (8.2) is the Jacobian matrix of the manipulator.

The dynamic model of the manipulator in the configuration space can be written in the form

$$\left.\begin{aligned} \dot{\mathbf{q}} &= \mathbf{v} \\ \mathbf{A}(\mathbf{q})\dot{\mathbf{v}} + \mathbf{b}(\mathbf{q}, \mathbf{v}) + \mathbf{g}(\mathbf{q}) &= \mathbf{T} \end{aligned}\right\} \tag{8.3}$$

where $\mathbf{q} = [q_1 \ q_2]^{\mathrm{T}}$ denotes the configuration vector of the joint angles, $\mathbf{v} = [v_1 \ v_2]^{\mathrm{T}}$ is the configuration space velocity vector, $\mathbf{A}(\mathbf{q}) \in \mathbb{R}^{2\times 2}$ stands for symmetric positive definite kinetic energy matrix, $\mathbf{b}(\mathbf{q}, \mathbf{v}) \in \mathbb{R}^{2\times 1}$ stands for the vector of Coriolis forces and centripetal forces, $\mathbf{g}(\mathbf{q}) \in \mathbb{R}^{2\times 1}$ is the vector of gravity terms, $\mathbf{T} \in \mathbb{R}^{2\times 1}$ is the vector of joint torques (sometimes it will be referred to $\mathbf{T}$ as the input force vector in the configuration space).

It was shown in [59] that elements of the matrix $\mathbf{A}$ can be calculated as

$$a_{11} = m_1 l_{c1}^2 + m_2 \left(l_1^2 + l_{c2}^2 + 2l_1 l_{c2} \cos q_2\right) + I_1 + I_2 \tag{8.4}$$

$$a_{12} = a_{21} = m_2 \left(l_{c2}^2 + l_1 l_{c2} \cos q_2\right) + I_2 \tag{8.5}$$

$$a_{22} = m_2 l_{c2}^2 + I_2. \tag{8.6}$$

Vector $\mathbf{b}(\mathbf{q}, \mathbf{v})$ can be obtained as

$$\mathbf{b}(\mathbf{q}, \mathbf{v}) = \mathbf{K}(\mathbf{q}, \mathbf{v})\mathbf{v} \tag{8.7}$$

where the matrix $\mathbf{K}(\mathbf{q}, \mathbf{v})$ is given as

$$\mathbf{K}(\mathbf{q}, \mathbf{v}) = \begin{bmatrix} hv_2 & hv_2 + hv_1 \\ -hv_1 & 0 \end{bmatrix}, \quad h = -m_2 l_1 l_{c2} \sin q_2. \quad (8.8)$$

If g stands for the magnitude of the gravitational acceleration which is directed along the negative y-axis in Figure 8.1, vector $\mathbf{g}(\mathbf{q})$ is calculated as

$$\mathbf{g}(\mathbf{q}) = \begin{bmatrix} (m_1 l_{c1} + m_2 l_1) g \cos q_1 + m_2 l_{c2} g \cos(q_1 + q_2) \\ m_2 l_{c2} g \cos(q_1 + q_2) \end{bmatrix}. \quad (8.9)$$

The presented dynamic model was used in MATLAB/Simulink for simulations of the mentioned tasks. The used manipulator's parameters are given in Table 8.1.

TABLE 8.1
Parameters of planar
manipulator.

Parameter	Value
m_1	1 kg
m_2	2 kg
l_1	1 m
l_2	1 m
l_{c1}	0.5 m
l_{c2}	0.5 m
I_1	0.087 kgm^2
I_2	0.180 kgm^2

8.1.2 Motion Synchronization Task in 2-D Space

The first task in which two planar manipulators will be considered is motion synchronization. Derivation of the control algorithm for the motion synchronization task is almost the same as the one given in Section 5.4. Both manipulators have identical parameters, given in Table 8.1. The goal in this task is to make positions of the manipulators (positions of their end-effectors, to be more precise) equal in the task space, and to have them track a reference trajectory. It is assumed that each manipulator's task space position is expressed in the x-y frame attached to its base. Let us assume that positions of the manipulators in the frames attached to their bases are given as $\mathbf{x}_1(\mathbf{q}_1) = [x_1 \; y_1]^{\mathrm{T}}$ and $\mathbf{x}_2(\mathbf{q}_2) = [x_2 \; y_2]^{\mathrm{T}}$, where $\mathbf{q}_1$ and $\mathbf{q}_2$ stand for the configuration vectors of the first and second manipulator, respectively. The reference trajectory is given as a two times differentiable vector-valued function of time $\mathbf{x}^{ref}(t) = [x^{ref}(t) \; y^{ref}(t)]^{\mathrm{T}}$. For the task to be executed, it is important to control two functions, denoted as: (i) synchronization $\boldsymbol{\varphi}_s = [\varphi_{s1} \; \varphi_{s2}]^{\mathrm{T}}$, and (ii) reference tracking $\boldsymbol{\varphi}_{rt} = [\varphi_{rt1} \; \varphi_{rt2}]^{\mathrm{T}}$, where each of them is a vector valued function, i.e., φ_s, $\varphi_{rt} \in \mathbb{R}^{2\times 1}$. These functions can be expressed in terms of manipulators' positions as

$$\varphi\left(\mathbf{q}_1, \mathbf{q}_2\right) = \begin{bmatrix} \varphi_s\left(\mathbf{q}_1, \mathbf{q}_2\right) \\ \varphi_{rt}\left(\mathbf{q}_1, \mathbf{q}_2\right) \end{bmatrix} = \begin{bmatrix} \mathbf{x}_1 - \mathbf{x}_2 \\ \mathbf{x}_1 + \mathbf{x}_2 \end{bmatrix}. \tag{8.10}$$

The task will be executed if both of these functions track their references $\varphi_s^{ref}(t)$ and $\varphi_{rt}^{ref}(t)$ defined as

$$\varphi^{ref}(t) = \begin{bmatrix} \varphi_s^{ref}(t) \\ \varphi_{rt}^{ref}(t) \end{bmatrix} = \begin{bmatrix} \mathbf{0} \\ 2\mathbf{x}^{ref} \end{bmatrix}. \tag{8.11}$$

The defined functions, φ_s and φ_{rt} depend on the configuration vectors $\mathbf{q}_1$ and $\mathbf{q}_2$ only. Thus, one can form the function vector, following the procedure given in Section 3.2.1, as follows

$$\mathbf{f} = \dot{\varphi}. \tag{8.12}$$

According to the procedure given in Section 4.1, the function vector reference is

$$\mathbf{f}^{ref} = \dot{\varphi}^{ref} - \mathbf{C}\left(\varphi - \varphi^{ref}\right) \tag{8.13}$$

where $\mathbf{C} \in \mathbb{R}^{4\times 4}$ is a constant diagonal matrix with positive diagonal entries

$$\mathbf{C} = \mathrm{diag}\left(c_1,\ c_2,\ c_3,\ c_4\right),\ c_i > 0,\ i = 1,\ 2,\ 3,\ 4. \tag{8.14}$$

The dynamics of the two manipulators can be written as

$$\left.\begin{aligned} \dot{\mathbf{q}}_1 &= \mathbf{v}_1 \\ \mathbf{A}_1(\mathbf{q}_1)\dot{\mathbf{v}}_1 + \mathbf{b}_1(\mathbf{q}_1, \mathbf{v}_1) + \mathbf{g}_1(\mathbf{q}_1) &= \mathbf{T}_1 \end{aligned}\right\} \tag{8.15}$$

$$\left.\begin{aligned} \dot{\mathbf{q}}_2 &= \mathbf{v}_2 \\ \mathbf{A}_2(\mathbf{q}_2)\dot{\mathbf{v}}_2 + \mathbf{b}_2(\mathbf{q}_2, \mathbf{v}_2) + \mathbf{g}_2(\mathbf{q}_2) &= \mathbf{T}_2 \end{aligned}\right\} \tag{8.16}$$

where subscripts 1 and 2 refer to the first and second manipulator. Components of the input force vectors in the configuration space, $\mathbf{T}_1$ and $\mathbf{T}_2$, are given as

$$\mathbf{T}_1 = \begin{bmatrix} T_{11} & T_{12} \end{bmatrix}^{\mathrm{T}},\ \mathbf{T}_2 = \begin{bmatrix} T_{21} & T_{22} \end{bmatrix}^{\mathrm{T}}. \tag{8.17}$$

In a shorter form, the dynamics of the system that consists of both manipulators can be written as

$$\left.\begin{aligned} \dot{\tilde{\mathbf{q}}} &= \tilde{\mathbf{v}} \\ \tilde{\mathbf{A}}(\tilde{\mathbf{q}})\dot{\tilde{\mathbf{v}}} + \tilde{\mathbf{b}}(\tilde{\mathbf{q}}, \tilde{\mathbf{v}}) + \tilde{\mathbf{g}}(\tilde{\mathbf{q}}) &= \tilde{\mathbf{T}}. \end{aligned}\right\} \tag{8.18}$$

The matrices and vectors that appear in (8.18) are given as

$$\begin{aligned} \tilde{\mathbf{A}} &= \begin{bmatrix} \mathbf{A}_1 & \mathbf{0}^{2\times 2} \\ \mathbf{0}^{2\times 2} & \mathbf{A}_2 \end{bmatrix},\ \tilde{\mathbf{q}} = \begin{bmatrix} \mathbf{q}_1 \\ \mathbf{q}_2 \end{bmatrix},\ \tilde{\mathbf{v}} = \begin{bmatrix} \mathbf{v}_1 \\ \mathbf{v}_2 \end{bmatrix},\ \tilde{\mathbf{b}} = \begin{bmatrix} \mathbf{b}_1 \\ \mathbf{b}_2 \end{bmatrix}, \\ \tilde{\mathbf{g}} &= \begin{bmatrix} \mathbf{g}_1 \\ \mathbf{g}_2 \end{bmatrix},\ \tilde{\mathbf{T}} = \begin{bmatrix} \mathbf{T}_1 \\ \mathbf{T}_2 \end{bmatrix}. \end{aligned} \tag{8.19}$$

Since $\tilde{\mathbf{A}} \in \mathbb{R}^{4 \times 4}$ is a nonsingular matrix, dynamics (8.18) can also be written in the following form

$$
\left.
\begin{aligned}
\dot{\tilde{\mathbf{q}}} &= \tilde{\mathbf{v}} \\
\dot{\tilde{\mathbf{v}}} &= \tilde{\mathbf{u}}_{\tilde{q}} - \tilde{\mathbf{A}}^{-1}\left(\tilde{\mathbf{b}} + \tilde{\mathbf{g}}\right), \quad \tilde{\mathbf{u}}_{\tilde{q}} = \tilde{\mathbf{A}}^{-1}\tilde{\mathbf{T}}.
\end{aligned}
\right\}
\tag{8.20}
$$

The defined function vector $\mathbf{f}$ can be expressed using the Jacobian matrices of the discussed manipulators, $\mathbf{J}_1 \in \mathbb{R}^{2 \times 2}$ and $\mathbf{J}_2 \in \mathbb{R}^{2 \times 2}$, as

$$
\mathbf{f} = \dot{\boldsymbol{\varphi}} = \begin{bmatrix} \dot{\boldsymbol{\varphi}}_s \\ \dot{\boldsymbol{\varphi}}_{rt} \end{bmatrix} = \underbrace{\begin{bmatrix} \mathbf{J}_1 & -\mathbf{J}_2 \\ \mathbf{J}_1 & \mathbf{J}_2 \end{bmatrix}}_{\mathbf{J}_f} \begin{bmatrix} \mathbf{v}_1 \\ \mathbf{v}_2 \end{bmatrix} = \mathbf{J}_f \tilde{\mathbf{v}}.
\tag{8.21}
$$

The matrix $\mathbf{J}_f \in \mathbb{R}^{4 \times 4}$ will be denoted as the function Jacobian matrix.

The tracking error vector $\mathbf{e} \in \mathbb{R}^{4 \times 1}$ for the given task can be expressed as

$$
\mathbf{e} = \begin{bmatrix} e_{xd} \\ e_{yd} \\ e_{xc} \\ e_{yc} \end{bmatrix} = \begin{bmatrix} x_1 - x_2 \\ y_1 - y_2 \\ x_1 + x_2 - 2x^{ref} \\ y_1 + y_2 - 2y^{ref} \end{bmatrix} = \begin{bmatrix} \boldsymbol{\varphi}_s \\ \boldsymbol{\varphi}_{rt} \end{bmatrix} - \begin{bmatrix} \boldsymbol{\varphi}_s^{ref} \\ \boldsymbol{\varphi}_{rt}^{ref} \end{bmatrix} = \boldsymbol{\varphi} - \boldsymbol{\varphi}^{ref}. \tag{8.22}
$$

The generalized error $\boldsymbol{\sigma} \in \mathbb{R}^{4 \times 1}$ is selected as

$$
\boldsymbol{\sigma} = \begin{bmatrix} \sigma_1 & \sigma_2 & \sigma_3 & \sigma_4 \end{bmatrix}^{\mathrm{T}} = \mathbf{f} - \mathbf{f}^{ref}. \tag{8.23}
$$

The control goal will be achieved if the system motion converges to the manifold $\boldsymbol{\sigma} = \mathbf{0}$.

The first-order dynamics of the generalized error is given as

$$
\dot{\boldsymbol{\sigma}} = \dot{\mathbf{f}} - \dot{\mathbf{f}}^{ref}. \tag{8.24}
$$

Using (8.21), (8.24) can be expressed as

$$
\dot{\boldsymbol{\sigma}} = \mathbf{J}_f \dot{\tilde{\mathbf{v}}} + \underbrace{\dot{\mathbf{J}}_f \tilde{\mathbf{v}}}_{\boldsymbol{\Upsilon}} - \dot{\mathbf{f}}^{ref}. \tag{8.25}
$$

Considering (8.20), (8.25) becomes

$$
\dot{\boldsymbol{\sigma}} = \mathbf{J}_f \tilde{\mathbf{u}}_{\tilde{q}} - \mathbf{J}_f \tilde{\mathbf{A}}^{-1}\left(\tilde{\mathbf{b}} + \tilde{\mathbf{g}}\right) + \boldsymbol{\Upsilon} - \dot{\mathbf{f}}^{ref}. \tag{8.26}
$$

Let us now introduce the control vector in the function space $\begin{bmatrix} u_{f1} & u_{f2} & u_{f3} & u_{f4} \end{bmatrix}^{\mathrm{T}} = \mathbf{u}_f \in \mathbb{R}^{4 \times 1}$, which is related to the control acceleration in the configuration space $\tilde{\mathbf{u}}_{\tilde{q}} \in \mathbb{R}^{4 \times 1}$ through the following equation

$$
\tilde{\mathbf{u}}_{\tilde{q}} = \mathbf{J}_f^{-1} \mathbf{u}_f. \tag{8.27}
$$

Therefore, it is assumed that $\mathbf{J}_f$ stays a nonsingular matrix for the entire time of operation. Now, dynamics (8.26) can be written as

$$\dot{\boldsymbol{\sigma}} = \mathbf{u}_f - \left[\mathbf{J}_f \tilde{\mathbf{A}}^{-1}\left(\tilde{\mathbf{b}} + \tilde{\mathbf{g}}\right) - \boldsymbol{\Upsilon} + \dot{\mathbf{f}}^{ref}\right]. \tag{8.28}$$

If the equivalent control is defined as

$$\mathbf{u}_f^{eq} = \mathbf{J}_f \tilde{\mathbf{A}}^{-1}\left(\tilde{\mathbf{b}} + \tilde{\mathbf{g}}\right) - \boldsymbol{\Upsilon} + \dot{\mathbf{f}}^{ref} \tag{8.29}$$

dynamics (8.28) can be written in a shorter form as

$$\dot{\boldsymbol{\sigma}} = \mathbf{u}_f - \mathbf{u}_f^{eq}. \tag{8.30}$$

Dynamics (8.30) is given in the form (3.142), for which control design process is described in Chapter 3. In this example, the control design will be the one based on the equivalent control estimation.

From (8.30), assuming that $\boldsymbol{\sigma}$ is available and with the equivalent control modeled as $\dot{\mathbf{u}}_f^{eq} = \mathbf{0}$, the equivalent control can be estimated as

$$\begin{aligned} \dot{\mathbf{z}} &= \mathbf{L}\left(\mathbf{u}_f - \mathbf{z} + \mathbf{L}\boldsymbol{\sigma}\right) \\ \hat{\mathbf{u}}_f^{eq} &= \mathbf{z} - \mathbf{L}\boldsymbol{\sigma} \end{aligned} \tag{8.31}$$

where $\mathbf{L} \in \mathbb{R}^{4 \times 4}$ is a constant gain matrix of the equivalent control observer given as

$$\mathbf{L} = \mathrm{diag}\left(l_1,\ l_2,\ l_3,\ l_4\right),\ l_i > 0,\ i = 1,\ 2,\ 3,\ 4 \tag{8.32}$$

while $\mathbf{u}_f^{eq} + \mathbf{L}\boldsymbol{\sigma} = \mathbf{z} \in \mathbb{R}^{4 \times 1}$ is the intermediate variable in the equivalent control estimation. The control vector in the function space is selected to enforce exponential convergence, and it is given as

$$\mathbf{u}_f = \hat{\mathbf{u}}_f^{eq} - \mathbf{D}\boldsymbol{\sigma} \tag{8.33}$$

with the constant diagonal matrix $\mathbf{D} \in \mathbb{R}^{4 \times 4}$

$$\mathbf{D} = \mathrm{diag}\left(d_1,\ d_2,\ d_3,\ d_4\right),\ d_i > 0,\ i = 1,\ 2,\ 3,\ 4. \tag{8.34}$$

After the control vector in the function space is selected, it is mapped back to the configuration space as follows

$$\tilde{\mathbf{T}} = \begin{bmatrix} \mathbf{T}_1 \\ \mathbf{T}_2 \end{bmatrix} = \tilde{\mathbf{A}}\mathbf{J}_f^{-1}\mathbf{u}_f. \tag{8.35}$$

The inverse of the function Jacobian matrix can be calculated using the inverse Jacobian matrices of the two manipulators as

$$\mathbf{J}_f^{-1} = \frac{1}{2}\begin{bmatrix} \mathbf{J}_1^{-1} & \mathbf{J}_1^{-1} \\ -\mathbf{J}_2^{-1} & \mathbf{J}_2^{-1} \end{bmatrix}. \tag{8.36}$$

In this way, it is possible to avoid calculation of the inverse matrix for a matrix of order four. Instead of that, inversion is calculated only for two Jacobian matrices of the manipulators whose order is equal to two.

The controlled system was simulated in MATLAB/Simulink for the reference trajectory given as

$$\mathbf{x}^{ref}(t) = \begin{bmatrix} x^{ref}(t) \\ y^{ref}(t) \end{bmatrix} = \begin{bmatrix} -0.2 + 0.2\sin(\pi t) \\ 1.45 + 0.2\cos(\pi t) \end{bmatrix}. \tag{8.37}$$

The initial joint angles of the two manipulators were

$$\mathbf{q}_1(0) = \begin{bmatrix} \frac{\pi}{3} \\ \frac{\pi}{4} \end{bmatrix}, \ \mathbf{q}_2(0) = \begin{bmatrix} \frac{\pi}{4} \\ \frac{\pi}{4} \end{bmatrix} \tag{8.38}$$

while the initial velocities were equal to zero. The matrices $\mathbf{C}$, $\mathbf{L}$, and $\mathbf{D}$ are given as

$$\begin{aligned} \mathbf{C} &= \operatorname{diag}(30, 30, 30, 30), \ \mathbf{L} = \operatorname{diag}(1200, 1200, 1200, 1200) \\ \mathbf{D} &= \operatorname{diag}(35, 35, 35, 35). \end{aligned} \tag{8.39}$$

To get realistic simulation results, all components of the control vector $\mathbf{u}_f$ were bounded and their minimum and maximum allowed values were -100 and 100 m/s^2, respectively. At the same time, the components of the configuration space input force vectors $\mathbf{T}_1$ and $\mathbf{T}_2$ were also bounded, with -250 and 250 N·m being their minimum and maximum allowed values.

Obtained responses are given in Figures 8.2–8.8. Both manipulators converge to the reference trajectory and tracking is achieved (see Figures 8.2 and 8.3). The manipulators move along a circular trajectory, as shown in Figure 8.4. All components of the tracking error vector exponentially converge to zero, meaning that the desired convergence is enforced (see Figure 8.5). Components of the function space control vector $\mathbf{u}_f$ reach their limit values -100 and 100 m/s^2 in the first 0.2 s of the simulation as can be noticed in Figure 8.6. However, input force vectors (joint torques) for the first and second manipulator are inside specified limits. Therefore, neither of the joint torques reaches -250 or 250 N·m (see Figures 8.7 and 8.8).

8.1.3   Object Manipulation Task in 2-D Space

The second task for which two planar manipulators were again employed was an object manipulation task. This task can be considered as a grasping and manipulation task. The task is very similar to that one discussed in Section 5.5, and the control algorithm derivation procedure given in this section is much like that one presented in Section 5.5. Within this task, it is important to control grasping force exerted on the object, and the position of the object in the task space. Such a task can be illustrated as in Figure 8.9. Each manipulator has an x-y frame attached to its base. Those two frames are marked with superscripts (1) and (2).

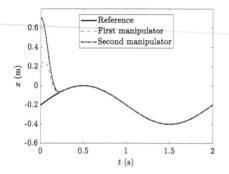

FIGURE 8.2
x-coordinate response.

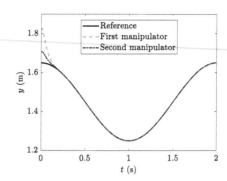

FIGURE 8.3
y-coordinate response.

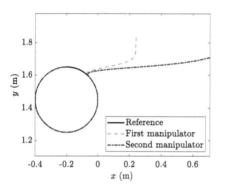

FIGURE 8.4
Trajectories of the manipulators in
the motion synchronization task.

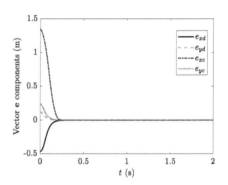

FIGURE 8.5
Vector **e** components in the motion
synchronization task.

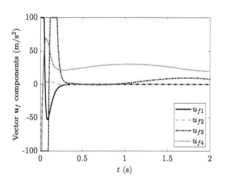

FIGURE 8.6
Vector $\mathbf{u}_f$ components in the motion
synchronization task.

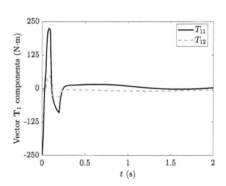

FIGURE 8.7
Vector $\mathbf{T}_1$ components in the motion
synchronization task.

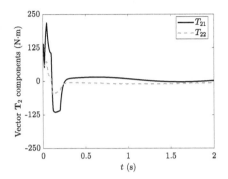

FIGURE 8.8
Vector $\mathbf{T}_2$ components in the motion synchronization task.

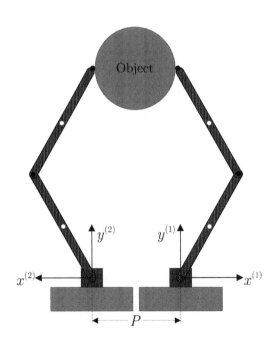

FIGURE 8.9
Two planar manipulators in an object manipulation task.

The task that should be executed can be described as follows. One has to control the x- and y-components of the grasping force applied to the object. The goal is to have the x-component of the grasping force at some specified value, and to bring the y-component to the zero value. This means that only existence of the x-component is desired. In addition, the position of the object, which can be described in one of two x-y frames, has to be controlled.

Since two x-y frames exist, it is important to see how coordinates from one frame can be expressed in another frame. Here, the following notation will be used. Symbol $x_j^{(i)}$ is the x-coordinate of the jth manipulator's end-effector expressed in the ith manipulator's frame. The same notation is valid for the y-coordinate. Let us suppose that the position of the second manipulator's end-effector in the frame attached to the second manipulator is $\left[x_2^{(2)} \; y_2^{(2)} \right]^{\mathrm{T}}$. If this position is to be expressed in the frame attached to the first manipulator, then it is given as

$$\begin{bmatrix} x_2^{(1)} \\ y_2^{(1)} \end{bmatrix} = \begin{bmatrix} -x_2^{(2)} - P \\ y_2^{(2)} \end{bmatrix} \tag{8.40}$$

where P is the distance between origins of the two frames.

There are four functions to be controlled in this system: (i) x-coordinate of the center of geometry for two end-effectors φ_{xc}, (ii) y-coordinate of the center of geometry for two end-effectors φ_{yc}, (iii) x-component of the grasping force φ_{gx}, and (iv) y-component of the grasping force φ_{gy}. For easier discussion, it is assumed that the object to be manipulated is initially placed in such a way that certain grasping force is applied to the object, meaning that contacts between the object and manipulators are established. The object dimensions and its initial placement with respect to the manipulators are assumed to be such that the grasping force exists as long as the distance between the end-effectors of the manipulators is less than P. The grasping force is modeled using a spring-damper model. Four scalar functions to be controlled are expressed as follows

$$\varphi\left(\mathbf{q}_1, \mathbf{q}_2, \mathbf{v}_1, \mathbf{v}_2\right) = \begin{bmatrix} \varphi_{xc}\left(\mathbf{q}_1, \mathbf{q}_2\right) \\ \varphi_{gy}\left(\mathbf{q}_1, \mathbf{q}_2, \mathbf{v}_1, \mathbf{v}_2\right) \\ \varphi_{gx}\left(\mathbf{q}_1, \mathbf{q}_2, \mathbf{v}_1, \mathbf{v}_2\right) \\ \varphi_{yc}\left(\mathbf{q}_1, \mathbf{q}_2\right) \end{bmatrix} =$$

$$= \begin{bmatrix} \frac{1}{2}\left(x_1^{(1)} + x_2^{(1)}\right) \\ K_e\left(y_1^{(1)} - y_2^{(2)}\right) + D_e\left(\dot{y}_1^{(1)} - \dot{y}_2^{(2)}\right) \\ K_e\left(-x_1^{(1)} - x_2^{(2)}\right) + D_e\left(-\dot{x}_1^{(1)} - \dot{x}_2^{(2)}\right) \\ \frac{1}{2}\left(y_1^{(1)} + y_2^{(1)}\right) \end{bmatrix} . \tag{8.41}$$

In (8.41), K_e and D_e are stiffness and damping coefficients characterizing the object in the contact points between the manipulators and object. The functions φ_{xc} and φ_{yc} are expressed in the frame attached to the first manipulator's base. Considering (8.40), the functions can also be written in the following form

$$\varphi\left(\mathbf{q}_1, \mathbf{q}_2, \mathbf{v}_1, \mathbf{v}_2\right) = \begin{bmatrix} \frac{1}{2}\left(x_1^{(1)} - x_2^{(2)} - P\right) \\ K_e\left(y_1^{(1)} - y_2^{(2)}\right) + D_e\left(\dot{y}_1^{(1)} - \dot{y}_2^{(2)}\right) \\ K_e\left(-x_1^{(1)} - x_2^{(2)}\right) + D_e\left(-\dot{x}_1^{(1)} - \dot{x}_2^{(2)}\right) \\ \frac{1}{2}\left(y_1^{(1)} + y_2^{(2)}\right) \end{bmatrix}. \tag{8.42}$$

Assuming that $x^{ref}(t)$ and $y^{ref}(t)$ are two times differentiable functions, and $F_g^{ref}(t)$ is a differentiable function, the references for the functions are given as

$$\varphi^{ref}(t) = \begin{bmatrix} \varphi_{xc}^{ref}(t) \\ \varphi_{gy}^{ref}(t) \\ \varphi_{gx}^{ref}(t) \\ \varphi_{yc}^{ref}(t) \end{bmatrix} = \begin{bmatrix} x^{ref} \\ 0 \\ F_g^{ref} \\ y^{ref} \end{bmatrix}. \tag{8.43}$$

The function vector will be defined, according the procedure given in Section 3.2.1, as follows

$$\mathbf{f} = \begin{bmatrix} f_{xc} \\ f_{gy} \\ f_{gx} \\ f_{yc} \end{bmatrix} = \begin{bmatrix} \dot{\varphi}_{xc} \\ \varphi_{gy} \\ \varphi_{gx} \\ \dot{\varphi}_{yc} \end{bmatrix} = \begin{bmatrix} \frac{1}{2}\left(\dot{x}_1^{(1)} - \dot{x}_2^{(2)}\right) \\ K_e\left(y_1^{(1)} - y_2^{(2)}\right) + D_e\left(\dot{y}_1^{(1)} - \dot{y}_2^{(2)}\right) \\ K_e\left(-x_1^{(1)} - x_2^{(2)}\right) + D_e\left(-\dot{x}_1^{(1)} - \dot{x}_2^{(2)}\right) \\ \frac{1}{2}\left(\dot{y}_1^{(1)} + \dot{y}_2^{(2)}\right) \end{bmatrix}. \tag{8.44}$$

Since $\left[\dot{x}_1^{(1)} \ \dot{y}_1^{(1)}\right]^{\mathrm{T}} = \mathbf{J}_1\mathbf{v}_1$ and $\left[\dot{x}_2^{(2)} \ \dot{y}_2^{(2)}\right] = \mathbf{J}_2\mathbf{v}_2$, one can write (8.44) in the following form

$$\mathbf{f} = \begin{bmatrix} \frac{1}{2} & 0 & 0 & 0 \\ 0 & D_e & 0 & 0 \\ 0 & 0 & -D_e & 0 \\ 0 & 0 & 0 & \frac{1}{2} \end{bmatrix} \begin{bmatrix} \mathbf{J}_1 & -\mathbf{J}_2 \\ \mathbf{J}_1 & \mathbf{J}_2 \end{bmatrix} \begin{bmatrix} \mathbf{v}_1 \\ \mathbf{v}_2 \end{bmatrix} + \begin{bmatrix} 0 \\ K_e\left(y_1^{(1)} - y_2^{(2)}\right) \\ -K_e\left(x_1^{(1)} + x_2^{(2)}\right) \\ 0 \end{bmatrix}. \tag{8.45}$$

According to the procedure given in Section 4.1, the reference for the function vector is

$$\mathbf{f}^{ref} = \begin{bmatrix} f_{xc}^{ref} \\ f_{gy}^{ref} \\ f_{gx}^{ref} \\ f_{yc}^{ref} \end{bmatrix} = \begin{bmatrix} \dot{\varphi}_{xc}^{ref} - c_1\left(\varphi_{xc} - \varphi_{xc}^{ref}\right) \\ \varphi_{gy}^{ref} \\ \varphi_{gx}^{ref} \\ \dot{\varphi}_{yc}^{ref} - c_4\left(\varphi_{yc} - \varphi_{yc}^{ref}\right) \end{bmatrix} \tag{8.46}$$

where c_1 and c_4 are positive constants.

Due to the grasping force applied to the object, there are reaction forces, and they introduce additional torques acting on the joints of both manipulators. These torques have to be included in the dynamic model of the manipulators. It can be done using the Jacobian matrices of the manipulators. The dynamics of the manipulators can be written as

$$\left.\begin{aligned} \dot{\mathbf{q}}_1 &= \mathbf{v}_1 \\ \mathbf{A}_1(\mathbf{q}_1)\dot{\mathbf{v}}_1 + \mathbf{b}_1(\mathbf{q}_1, \mathbf{v}_1) + \mathbf{g}_1(\mathbf{q}_1) &= \mathbf{T}_1 + \mathbf{T}_1^r \end{aligned}\right\} \qquad (8.47)$$

$$\left.\begin{aligned} \dot{\mathbf{q}}_2 &= \mathbf{v}_2 \\ \mathbf{A}_2(\mathbf{q}_2)\dot{\mathbf{v}}_2 + \mathbf{b}_2(\mathbf{q}_2, \mathbf{v}_2) + \mathbf{g}_2(\mathbf{q}_2) &= \mathbf{T}_2 + \mathbf{T}_2^r \end{aligned}\right\} \qquad (8.48)$$

where torques appearing due to the reaction forces, $\mathbf{T}_1^r \in \mathbb{R}^{2 \times 1}$ and $\mathbf{T}_2^r \in \mathbb{R}^{2 \times 1}$, are given by

$$\begin{aligned} \mathbf{T}_1^r &= \mathbf{J}_1^T \begin{bmatrix} \varphi_{gx} \\ \varphi_{gy} \end{bmatrix} \\ \mathbf{T}_2^r &= \mathbf{J}_2^T \begin{bmatrix} \varphi_{gx} \\ -\varphi_{gy} \end{bmatrix}. \end{aligned} \qquad (8.49)$$

Now, the dynamics of the system consisting of both manipulators is

$$\left.\begin{aligned} \dot{\tilde{\mathbf{q}}} &= \tilde{\mathbf{v}} \\ \tilde{\mathbf{A}}(\tilde{\mathbf{q}})\dot{\tilde{\mathbf{v}}} + \tilde{\mathbf{b}}(\tilde{\mathbf{q}}, \tilde{\mathbf{v}}) + \tilde{\mathbf{g}}(\tilde{\mathbf{q}}) - \tilde{\mathbf{T}}^r &= \tilde{\mathbf{T}}. \end{aligned}\right\} \qquad (8.50)$$

The matrices and vectors appearing in (8.50) are given in (8.19) and by the following equation

$$\tilde{\mathbf{T}}^r = \begin{bmatrix} \mathbf{T}_1^r \\ \mathbf{T}_2^r \end{bmatrix} \qquad (8.51)$$

Using the fact that $\tilde{\mathbf{A}} \in \mathbb{R}^{4 \times 4}$ is a nonsingular matrix, an alternative form of (8.50) is

$$\left.\begin{aligned} \dot{\tilde{\mathbf{q}}} &= \tilde{\mathbf{v}} \\ \dot{\tilde{\mathbf{v}}} &= \tilde{\mathbf{u}}_{\tilde{q}} - \tilde{\mathbf{A}}^{-1}\left(\tilde{\mathbf{b}} + \tilde{\mathbf{g}} - \tilde{\mathbf{T}}^r\right), \quad \tilde{\mathbf{u}}_{\tilde{q}} = \tilde{\mathbf{A}}^{-1}\tilde{\mathbf{T}}. \end{aligned}\right\} \qquad (8.52)$$

The tracking error vector $\mathbf{e} \in \mathbb{R}^{4 \times 1}$ and the generalized error $\boldsymbol{\sigma} \in \mathbb{R}^{4 \times 1}$ are defined as

$$\mathbf{e} = \begin{bmatrix} e_{xc} \\ e_{gy} \\ e_{gx} \\ e_{yc} \end{bmatrix} = \boldsymbol{\varphi} - \boldsymbol{\varphi}^{ref} \qquad (8.53)$$

$$\boldsymbol{\sigma} = \begin{bmatrix} \sigma_1 \\ \sigma_2 \\ \sigma_3 \\ \sigma_4 \end{bmatrix} = \mathbf{f} - \mathbf{f}^{ref}. \qquad (8.54)$$

The control goal will be achieved if the system motion converges to the manifold $\boldsymbol{\sigma} = \mathbf{0}$.

The first-order dynamics of the generalized error can be written as

$$\dot{\boldsymbol{\sigma}} = \dot{\mathbf{f}} - \dot{\mathbf{f}}^{ref} \qquad (8.55)$$

Considering (8.45), (8.55) can further be written as

$$\dot{\boldsymbol{\sigma}} = \begin{bmatrix} \frac{1}{2} & 0 & 0 & 0 \\ 0 & D_e & 0 & 0 \\ 0 & 0 & -D_e & 0 \\ 0 & 0 & 0 & \frac{1}{2} \end{bmatrix} \begin{bmatrix} \mathbf{J}_1 & -\mathbf{J}_2 \\ \mathbf{J}_1 & \mathbf{J}_2 \end{bmatrix} \begin{bmatrix} \dot{\mathbf{v}}_1 \\ \dot{\mathbf{v}}_2 \end{bmatrix} +$$

$$+ \begin{bmatrix} \frac{1}{2} & 0 & 0 & 0 \\ 0 & D_e & 0 & 0 \\ 0 & 0 & -D_e & 0 \\ 0 & 0 & 0 & \frac{1}{2} \end{bmatrix} \begin{bmatrix} \dot{\mathbf{J}}_1 & -\dot{\mathbf{J}}_2 \\ \dot{\mathbf{J}}_1 & \dot{\mathbf{J}}_2 \end{bmatrix} \begin{bmatrix} \mathbf{v}_1 \\ \mathbf{v}_2 \end{bmatrix} + \begin{bmatrix} 0 \\ K_e \left(\dot{y}_1^{(1)} - \dot{y}_2^{(2)} \right) \\ -K_e \left(\dot{x}_1^{(1)} + \dot{x}_2^{(2)} \right) \\ 0 \end{bmatrix} - \dot{\mathbf{f}}^{ref}.$$

$$(8.56)$$

The dynamics (8.56) can as well be rewritten in the already introduced standard form

$$\dot{\boldsymbol{\sigma}} = \mathbf{J}_f \dot{\mathbf{v}} + \boldsymbol{\Upsilon} - \dot{\mathbf{f}}^{ref} \tag{8.57}$$

where the function Jacobian matrix $\mathbf{J}_f$ and $\boldsymbol{\Upsilon}$ are

$$\mathbf{J}_f = \begin{bmatrix} \frac{1}{2} & 0 & 0 & 0 \\ 0 & D_e & 0 & 0 \\ 0 & 0 & -D_e & 0 \\ 0 & 0 & 0 & \frac{1}{2} \end{bmatrix} \begin{bmatrix} \mathbf{J}_1 & -\mathbf{J}_2 \\ \mathbf{J}_1 & \mathbf{J}_2 \end{bmatrix} \tag{8.58}$$

$$\boldsymbol{\Upsilon} = \begin{bmatrix} \frac{1}{2} & 0 & 0 & 0 \\ 0 & D_e & 0 & 0 \\ 0 & 0 & -D_e & 0 \\ 0 & 0 & 0 & \frac{1}{2} \end{bmatrix} \begin{bmatrix} \dot{\mathbf{J}}_1 & -\dot{\mathbf{J}}_2 \\ \dot{\mathbf{J}}_1 & \dot{\mathbf{J}}_2 \end{bmatrix} \begin{bmatrix} \mathbf{v}_1 \\ \mathbf{v}_2 \end{bmatrix} + \begin{bmatrix} 0 \\ K_e \left(\dot{y}_1^{(1)} - \dot{y}_2^{(2)} \right) \\ -K_e \left(\dot{x}_1^{(1)} + \dot{x}_2^{(2)} \right) \\ 0 \end{bmatrix}. \tag{8.59}$$

Considering (8.52), (8.57) becomes

$$\dot{\boldsymbol{\sigma}} = \mathbf{J}_f \left[\tilde{\mathbf{u}}_{\tilde{q}} - \tilde{\mathbf{A}}^{-1} \left(\tilde{\mathbf{b}} + \tilde{\mathbf{g}} - \tilde{\mathbf{T}}^r \right) \right] + \boldsymbol{\Upsilon} - \dot{\mathbf{f}}^{ref}. \tag{8.60}$$

Let us now introduce the control vector in the function space $[u_{f1} \ u_{f2} \ u_{f3} \ u_{f4}]^{\mathrm{T}} = \mathbf{u}_f \in \mathbb{R}^{4 \times 1}$, which is related with the control acceleration vector in the configuration space as follows

$$\tilde{\mathbf{u}}_{\tilde{q}} = \mathbf{J}_f^{-1} \mathbf{u}_f. \tag{8.61}$$

It is again assumed that $\mathbf{J}_f$ stays a nonsingular matrix for the entire time of operation. Dynamics (8.60) becomes

$$\dot{\boldsymbol{\sigma}} = \mathbf{u}_f - \mathbf{J}_f \tilde{\mathbf{A}}^{-1} \left(\tilde{\mathbf{b}} + \tilde{\mathbf{g}} - \tilde{\mathbf{T}}^r \right) + \boldsymbol{\Upsilon} - \dot{\mathbf{f}}^{ref}. \tag{8.62}$$

If the equivalent control is defined as

$$\mathbf{u}_f^{eq} = \mathbf{J}_f \tilde{\mathbf{A}}^{-1} \left(\tilde{\mathbf{b}} + \tilde{\mathbf{g}} - \tilde{\mathbf{T}}^r \right) - \boldsymbol{\Upsilon} + \dot{\mathbf{f}}^{ref} \tag{8.63}$$

the first-order dynamics of the generalized error can finally be written as

$$\dot{\boldsymbol{\sigma}} = \mathbf{u}_f - \mathbf{u}_f^{eq}. \tag{8.64}$$

If one compares (8.30) and (8.64), it can be concluded that two different tasks can be described in the same framework. Therefore, the control synthesis for this task is identical as for the previous. If $\boldsymbol{\sigma}$ is available and equivalent control is modeled as $\dot{\mathbf{u}}_f^{eq} = \mathbf{0}$ (which was also assumed in the previous task), the equivalent control can be estimated as in (8.31), and control which will enforce exponential convergence to the manifold $\boldsymbol{\sigma} = \mathbf{0}$ can be chosen as in (8.33). The matrices $\mathbf{L} \in \mathbb{R}^{4 \times 4}$ and $\mathbf{D} \in \mathbb{R}^{4 \times 4}$ have the same forms as in the previous task, and they are expressed in (8.32) and (8.34). Once the control vector $\mathbf{u}_f$ is selected in the function space, it has to be transformed back to the configuration space. This transformation is the same as for the previous task, and it is given in (8.35). The only difference is that the function Jacobian is different for two tasks. For this task, the inverse of the function Jacobian can be calculated as

$$\mathbf{J}_f^{-1} = \frac{1}{2} \begin{bmatrix} \mathbf{J}_1^{-1} & \mathbf{J}_1^{-1} \\ -\mathbf{J}_2^{-1} & \mathbf{J}_2^{-1} \end{bmatrix} \begin{bmatrix} 2 & 0 & 0 & 0 \\ 0 & D_e^{-1} & 0 & 0 \\ 0 & 0 & -D_e^{-1} & 0 \\ 0 & 0 & 0 & 2 \end{bmatrix}. \tag{8.65}$$

The controlled system was simulated in MATLAB/Simulink. The references for the controlled functions were

$$\boldsymbol{\varphi}^{ref}(t) = \begin{bmatrix} \varphi_{xc}^{ref}(t) \\ \varphi_{gy}^{ref}(t) \\ \varphi_{gx}^{ref}(t) \\ \varphi_{yc}^{ref}(t) \end{bmatrix} = \begin{bmatrix} 0.05 + 0.2 \sin(\pi t) \\ 0 \\ 18 + 10 \sin(\pi t) \\ 1.2 + 0.2 \cos(\pi t) \end{bmatrix}. \tag{8.66}$$

The initial joint angles of the two manipulators were

$$\mathbf{q}_1(0) = \begin{bmatrix} \frac{\pi}{3} + \frac{\pi}{100} \\ \frac{\pi}{3} \end{bmatrix}, \quad \mathbf{q}_2(0) = \begin{bmatrix} \frac{\pi}{3} + \frac{\pi}{100} \\ \frac{\pi}{3} \end{bmatrix} \tag{8.67}$$

while the initial velocities were equal to zero. The distance between two frames was $P = 0.4$ m, while the stiffness and damping coefficient characterizing the object in the contact points between the manipulators and object were $K_e = 250$ kg/s^2, $D_e = 5$ kg/s. The constants used in generalized error formulation are given by

$$c_1 = c_4 = 30 \tag{8.68}$$

and the matrices $\mathbf{L}$ and $\mathbf{D}$ were

$$\mathbf{L} = \text{diag}\,(1200,\ 1200,\ 1200,\ 1200)\,, \quad \mathbf{D} = \text{diag}\,(35,\ 35,\ 35,\ 35)\,. \tag{8.69}$$

In order to have realistic simulation results, all four components of the control vector $\mathbf{u}_f$ were bounded so that their absolute values cannot exceed

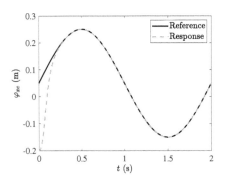

FIGURE 8.10

φ_{xc} function response.

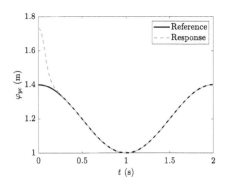

FIGURE 8.11

φ_{yc} function response.

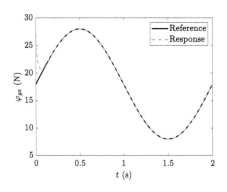

FIGURE 8.12

φ_{gx} function response.

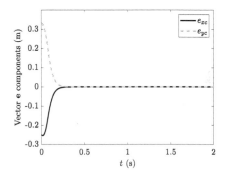

FIGURE 8.13

Tracking errors for the φ_{xc} and φ_{yc} functions.

50 m/s^2 for u_{f1} and u_{f4}, while that limit was 500 N/s for u_{f2} and u_{f3}. At the same time, components of the configuration space input force vectors $\mathbf{T}_1$ and $\mathbf{T}_2$ were also bounded, and -250 and 250 N·m were their minimum and maximum allowed values.

The simulation results are shown in Figures 8.10–8.18. The functions φ_{xc}, φ_{yc}, and φ_{gx} converge to their references and tracking is achieved (see Figures 8.10–8.12). The responses of the tracking errors are as desired, since they are all exponentially converging to zero (see Figures 8.13 and 8.14). All components of the function space control vector $\mathbf{u}_f$ except u_{f2} reach their saturation values at the beginning of the simulation (see Figures 8.15 and 8.16). However, in the configuration space, only the first joint torque of the first manipulator reaches its saturation value, since it is equal to -250 N·m for a short time interval at the beginning of the simulation, as can be seen from Figures 8.17 and 8.18. The responses show that the object manipulation task is successfully executed.

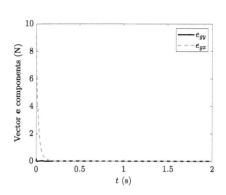

FIGURE 8.14
Tracking error for the φ_{gy} and φ_{gx} functions.

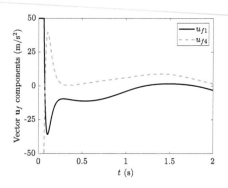

FIGURE 8.15
The first component and fourth component of the vector $\mathbf{u}_f$ in the object manipulation task in 2-D space.

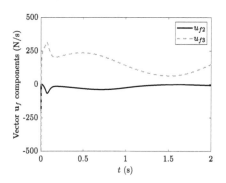

FIGURE 8.16
The second component and third component of the vector $\mathbf{u}_f$ in the object manipulation task in 2-D space.

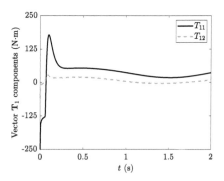

FIGURE 8.17
Vector $\mathbf{T}_1$ components in the object manipulation task in 2-D space.

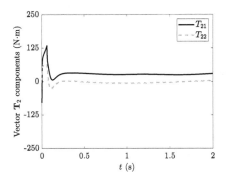

FIGURE 8.18

Vector $\mathbf{T}_2$ components in the object manipulation task in 2-D space.

8.2 Bilateral System

In this section, control in a bilateral system will be considered. A bilateral control system enables interaction between a human operator and a remote environment, which is established using interface devices. In such systems, the operator acts on the environment by two functionally related systems (devices). Contact is established between the operator and one device (master device), while the other device (slave device) and environment are in contact. The operator defines the task for the slave device by manipulating the master device. The master device acquires motion data from the operator, and, additionally, exerts force on the operator. The slave device reproduces the operator motion at the remote location, it senses interactions with the environment, and also transmits information on its motion and interaction force to the master device and operator. The described functional relationship is known in the literature as a bilateral system. Bilateral control systems have been extensively discussed in [54].

The ideal behavior of the bilateral control system can be defined as follows [86]. The position responses on the master side and on the slave side are identical, and, additionally, the force responses that are produced on the master side and on the slave side by the operator's input are also identical, whatever the dynamics of the object being manipulated is. Loosely speaking, the bilateral system enables the operator to operate an object or execute a task from a distance, while, at the same time, he can sense the force generated on the slave side. Therefore, the operator feels that he himself is operating the object located on the slave side. This kind of operation is defined as 'ideal kinesthetic coupling'. However, one has to take care while understanding the provided definition from the point of view of the ability of systems to control the position and the force concurrently.

Four players exist in a bilateral system: (i) operator, (ii) master side device (robot), (iii) slave side device (robot), and (iv) object to be manipulated. They all have their own dynamical properties. The inputs to the system can be identified as the operator's force and motion and also the force generated because of interaction of the slave side device and the object to be manipulated.

In this section, both the master device and slave device are assumed 1-DOF robotic manipulators, and they are described in (8.70) and (8.71). The operator's motion is given in (8.72), while the interaction force between operator and the master device is modeled as in (8.73). The operator is modeled as a full 1-DOF mechanical system whose inputs are the operator force T_{op} and the interaction force with master device F_h. In this study, interaction between the operator and master device is modeled as a spring-damper system (8.73). In addition, the model of the interaction force between the environment and slave device F_s is given in (8.74).

$$\left.\begin{array}{c} \dot{q}_m = v_m \\ a_m\left(q_m\right)\dot{v}_m + b_m\left(q_m, v_m\right) + g_m\left(q_m\right) = T_m - T_{mext} + F_h \end{array}\right\} \quad (8.70)$$

$$\left.\begin{array}{c} \dot{q}_s = v_s \\ a_s\left(q_s\right)\dot{v}_s + b_s\left(q_s, v_s\right) + g_s\left(q_s\right) = T_s - T_{sext} - F_s \end{array}\right\} \quad (8.71)$$

$$\left.\begin{array}{c} \dot{q}_h = v_h \\ a_h\left(q_h\right)\dot{v}_h + b_h\left(q_h, v_h\right) + g_h\left(q_h\right) = T_{op} - F_h \end{array}\right\} \quad (8.72)$$

$$D_h\left(v_h - v_m\right) + K_h\left(q_h - q_m\right) = F_h \quad (8.73)$$

$$D_e\left(v_s - v_e\right) + K_e\left(q_s - q_e\right) = F_s. \quad (8.74)$$

Models (8.70)–(8.74) are taken from [54]. In (8.70)–(8.74), the subscript m stands for parameters and variables related to the master side, the subscript s denotes slave side parameters and variables, the subscript h is related to the human operator, the subscript e represents environment parameters and variables, while T_{ext} represents the external force on the master and slave sides. The given model describes a bilateral system in the configuration space.

The master (8.70) and the slave devices (8.71) are active systems that have T_m and T_s as the control forces, while F_h and F_s represent the interaction forces on the master and slave side, respectively. By assumption, the slave device is in contact with the environment moving on a trajectory defined by position q_e and velocity v_e. Interaction force F_s exists only if there exists contact between the slave device and environment. In this discussion, it will be assumed $F_s \geq 0$. Parameters D_e and K_e define the properties of the object (environment) in the contact point. The environment motion is external input to the system that is assumed to be unknown.

The force exerted by the operator is dependent on the motion of the operator and also on the master device motion. If it is assumed that the properties of the operator in the contact point are given by D_h, K_h, while the motion is defined by position q_h and velocity v_h, then the force due to motion of the master device is defined as in (8.73). The position being dictated by the operator and position of the environment are treated in the same way. Similarly as for the force F_s, the force F_h exists only if contact is established. It is assumed that the contact is such that both push and pull forces can exist on the master side.

By assumption, the forces and positions on the master side and slave side are measured. The dynamics given in (8.70)–(8.74) represents four systems in interaction.

Control has to be designed to preserve bilateral relations between two systems, the human operator-master system and the slave-environment system. Input force coming from the operator, T_{op}, is basically an active input to the system. However, it cannot be changed by the system components; thus, T_{op} acts as a reference. By assumption, the operator himself responds on the occurrence of the interaction force on the slave side by a change in its position; thus, the magnitude of the interaction force and motion of the slave device are adjusted. The master device has two roles: (i) definition of the position reference for the slave device and (ii) generation of the force F_h that is equal to the interaction force on the slave side. In an ideal scenario, the force F_h opposes the operator's motion in the same way as the force of a direct touch on the operator's hand with a manipulated object would do. Thus, three active command inputs exist in the system (T_{op}, T_m, T_s), but T_{op} is considered as an input that cannot be changed by the control system. Therefore, T_{op} governs motion on the master side, and ideally, if no interaction force exists on the slave side, F_h is equal to zero and the operator position, master device position and slave device position are all the same. Control inputs on the master and slave side are forces T_m and T_s, respectively. Due to the interaction between the environment and slave device, the interaction force F_s appears. The general structure of the bilateral system is shown in Figure 8.19.

As shown in Figure 8.19, the master device acts as an environment for the human operator; the force due to the interaction between the master device and operator is F_h. The master system is impressing force F_h on the operator. The force being extended to the operator is equal to the force created by the relative motion of the master device and operator. That force can be controlled to be equal to the interaction force on the slave side, which occurs due to the interaction with the environment, as given in (8.74). Therefore, if $F_s = F_h$, the operator will sense a force equal to the interaction force on the slave side and thus he will have a sense of direct touch.

The bilateral controller has to calculate the control forces for master and slave systems, on the basis of the known positions of the systems (q_m and q_s), the interaction forces between the master device and operator F_h and slave device and environment F_s.

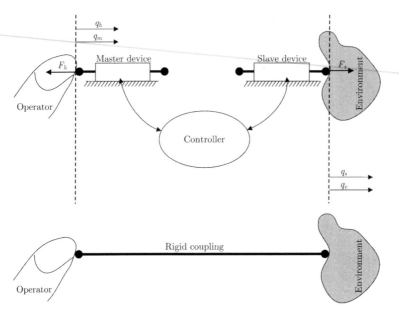

FIGURE 8.19
Structure of a bilateral system as it is described in (8.70)–(8.74).

In the discussed bilateral system, there are two functions to be controlled:
(i) position synchronization φ_{ps} and (ii) force synchronization φ_{fs}. The functions are defined as follows

$$\varphi\left(q_m, q_s, F_s, F_h\right) = \begin{bmatrix} \varphi_{ps}\left(q_m, q_s\right) \\ \varphi_{fs}\left(F_s, F_h\right) \end{bmatrix} = \begin{bmatrix} q_m - q_s \\ F_s - F_h \end{bmatrix} \tag{8.75}$$

while their references are

$$\varphi^{ref}\left(t\right) = \begin{bmatrix} \varphi_{ps}^{ref}\left(t\right) \\ \varphi_{fs}^{ref}\left(t\right) \end{bmatrix} = \begin{bmatrix} 0 \\ 0 \end{bmatrix}. \tag{8.76}$$

Since, the interaction forces F_s and F_h depend on the slave and master velocities, the function vector will be defined, according the procedure given in Section 3.2.1, as follows

$$
\begin{aligned}
\mathbf{f} &= \begin{bmatrix} f_{ps} \\ f_{fs} \end{bmatrix} = \begin{bmatrix} \dot{\varphi}_{ps} \\ \varphi_{fs} \end{bmatrix} = \begin{bmatrix} \dot{q}_m - \dot{q}_s \\ F_s - F_h \end{bmatrix} = \\
&= \begin{bmatrix} v_m - v_s \\ D_e\left(v_s - v_e\right) + K_e\left(q_s - q_e\right) - D_h\left(v_h - v_m\right) - K_h\left(q_h - q_m\right) \end{bmatrix}.
\end{aligned}
\tag{8.77}
$$

According to the procedure given in Section 4.1, the reference for the function vector is

$$\mathbf{f}^{ref} = \begin{bmatrix} f_{ps}^{ref} \\ f_{fs}^{ref} \end{bmatrix} = \begin{bmatrix} \dot{\varphi}_{ps}^{ref} - c_1\left(\varphi_{ps} - \varphi_{ps}^{ref}\right) \\ \varphi_{fs}^{ref} \end{bmatrix} \tag{8.78}$$

where c_1 is a positive constant.

In this section, it will be illustrated how a control system can be organized in layers, where each layer executes a part of the total control action. Therefore, it will be assumed that the master and slave devices have locally attached disturbance observers which are compensating disturbances acting in the configuration space.

If one assumes that inertia of the master device is given as $a_m(q_m) = a_{mn} + \Delta a_m(q_m)$ where only nominal inertia a_{mn} is known, and $\Delta a_m(q_m)$ is the unknown variation, the dynamics of the master device can be written as

$$\left.\begin{array}{c} \dot{q}_m = v_m \\ a_{mn}\dot{v}_m = T_m - \underbrace{(\Delta a_m \dot{v}_m + b_m + g_m + T_{mext} - F_h)}_{T_{mdis}} = T_m - T_{mdis}. \end{array}\right\} \quad (8.79)$$

On the other hand, if only nominal inertia of the slave device a_{sn} is known and its total inertia is $a_s(q_s) = a_{sn} + \Delta a_s(q_s)$, with $\Delta a_s(q_s)$ being the unknown variation, then the dynamics of the slave device can be rewritten as

$$\left.\begin{array}{c} \dot{q}_s = v_s \\ a_{sn}\dot{v}_s = T_s - \underbrace{(\Delta a_s \dot{v}_s + b_s + g_s + T_{sext} + F_s)}_{T_{sdis}} = T_s - T_{sdis}. \end{array}\right\} \quad (8.80)$$

In (8.79) and (8.80), T_{mdis} is the generalized disturbance acting on the master device, while the generalized disturbance acting on the slave device is denoted as T_{sdis}.

If velocities of the master and slave device are available (measured or calculated), and with acting generalized disturbances on the master and slave devices modeled as $\dot{T}_{mdis} = 0$ and $\dot{T}_{sdis} = 0$, the disturbances can be estimated as follows

$$\left.\begin{array}{rcl} \dot{z}_m & = & \frac{l_m}{a_{mn}}(T_m - z_m + l_m v_m) \\ \hat{T}_{mdis} & = & z_m - l_m v_m \end{array}\right. \quad (8.81)$$

$$\left.\begin{array}{rcl} \dot{z}_s & = & \frac{l_s}{a_{sn}}(T_s - z_s + l_s v_s) \\ \hat{T}_{sdis} & = & z_s - l_s v_m \end{array}\right. \quad (8.82)$$

where l_m and l_s are positive constant gains, while $z_m = T_{mdis} + l_m v_m$ and $z_s = T_{sdis} + l_s v_s$ are the intermediate variables in the estimation of the disturbances. Now, it will be assumed that both T_m and T_s consist of the estimated corresponding generalized disturbance and an additional term which is to be calculated by a high-level controller, i.e., the forces can be expressed as

$$\left.\begin{array}{rcl} T_m & = & \hat{T}_{mdis} + T_{mc} \\ T_s & = & \hat{T}_{sdis} + T_{sc}. \end{array}\right. \quad (8.83)$$

Taking (8.83) into account, (8.79) and (8.80) become

$$\left.\begin{array}{c} \dot{q}_m = v_m \\ a_{mn}\dot{v}_m = T_{mc} - \underbrace{\left(T_{mdis} - \hat{T}_{mdis}\right)}_{p_m} = T_{mc} - p_m \end{array}\right\} \quad (8.84)$$

$$\dot{q}_s = v_s$$
$$a_{sn}\dot{v}_s = T_{sc} - \underbrace{\left(T_{sdis} - \hat{T}_{sdis}\right)}_{p_s} = T_{sc} - p_s. \left.\vphantom{\begin{matrix}1\\1\\1\end{matrix}}\right\} \tag{8.85}$$

In (8.84) and (8.85), p_m and p_s are errors appearing in the estimation of the disturbance forces on the master and slave side, respectively. Dynamics (8.84) and (8.85) can be combined and written in a compact form as

$$\dot{\mathbf{q}} = \mathbf{v}$$
$$\mathbf{A}_n\dot{\mathbf{q}} = \mathbf{T}_c - \underbrace{\left(\mathbf{T}_{dis} - \hat{\mathbf{T}}_{dis}\right)}_{\mathbf{p}} = \mathbf{T}_c - \mathbf{p} \left.\vphantom{\begin{matrix}1\\1\\1\end{matrix}}\right\} \tag{8.86}$$

where

$$\begin{aligned}
\mathbf{A}_n &= \begin{bmatrix} a_{mn} & 0 \\ 0 & a_{sn} \end{bmatrix}, \ \mathbf{q} = \begin{bmatrix} q_m \\ q_s \end{bmatrix}, \ \mathbf{v} = \begin{bmatrix} v_m \\ v_s \end{bmatrix}, \ \mathbf{T}_c = \begin{bmatrix} T_{mc} \\ T_{sc} \end{bmatrix}, \\
\mathbf{T}_{dis} &= \begin{bmatrix} T_{mdis} \\ T_{sdis} \end{bmatrix}, \ \mathbf{p} = \begin{bmatrix} p_m \\ p_s \end{bmatrix}.
\end{aligned} \tag{8.87}$$

Since $\mathbf{A}_n$ is a nonsingular matrix, dynamics (8.86) can also be written in the following form

$$\dot{\mathbf{q}} = \mathbf{v}$$
$$\dot{\mathbf{v}} = \mathbf{u}_{qc} - \mathbf{A}_n^{-1}\mathbf{p}, \ \mathbf{u}_{qc} = \mathbf{A}_n^{-1}\mathbf{T}_c. \left.\vphantom{\begin{matrix}1\\1\end{matrix}}\right\} \tag{8.88}$$

Considering the defined functions for this task and their references, the tracking error vector $\mathbf{e} \in \mathbb{R}^{2 \times 1}$ is defined as

$$\mathbf{e} = \begin{bmatrix} e_{ps} \\ e_{fs} \end{bmatrix} = \begin{bmatrix} q_m - q_s \\ F_s - F_h \end{bmatrix} = \boldsymbol{\varphi} - \boldsymbol{\varphi}^{ref}. \tag{8.89}$$

The generalized error is given as

$$\boldsymbol{\sigma} = \begin{bmatrix} \sigma_1 \\ \sigma_2 \end{bmatrix} = \mathbf{f} - \mathbf{f}^{ref}. \tag{8.90}$$

The goal of the control system is to enforce convergence of the motion of the bilateral system to the manifold $\boldsymbol{\sigma} = \mathbf{0}$.

The first-order dynamics of the generalized error is

$$\dot{\boldsymbol{\sigma}} = \dot{\mathbf{f}} - \dot{\mathbf{f}}^{ref}. \tag{8.91}$$

Taking into account (8.77), (8.91) can further be written as

$$\dot{\boldsymbol{\sigma}} = \begin{bmatrix} 1 & -1 \\ D_h & D_e \end{bmatrix} \begin{bmatrix} \dot{v}_m \\ \dot{v}_s \end{bmatrix} + \begin{bmatrix} 0 \\ -D_e\dot{v}_e + K_e\left(v_s - v_e\right) - D_h\dot{v}_h + K_h\left(v_m - v_h\right) \end{bmatrix} - \dot{\mathbf{f}}^{ref} \tag{8.92}$$

The last equation can be rewritten in the already introduced standard form as

$$\dot{\boldsymbol{\sigma}} = \mathbf{J}_f\dot{\mathbf{v}} + \boldsymbol{\Upsilon} - \dot{\mathbf{f}}^{ref} \tag{8.93}$$

where

$$\mathbf{J}_f = \begin{bmatrix} 1 & -1 \\ D_h & D_e \end{bmatrix} \tag{8.94}$$

$$\boldsymbol{\Upsilon} = \begin{bmatrix} 0 \\ -D_e \dot{v}_e + K_e (v_s - v_e) - D_h \dot{v}_h + K_h (v_m - v_h) \end{bmatrix}. \tag{8.95}$$

Considering (8.88), (8.93) becomes

$$\dot{\boldsymbol{\sigma}} = \mathbf{J}_f \mathbf{u}_{qc} - \mathbf{J}_f \mathbf{A}_n^{-1} \mathbf{p} + \boldsymbol{\Upsilon} - \dot{\mathbf{f}}^{ref} \tag{8.96}$$

Let us now introduce the control vector in the function space $[u_{f1} \ u_{f2}]^T = \mathbf{u}_f \in \mathbb{R}^{2 \times 1}$, which is related to the control acceleration vector $\mathbf{u}_{qc} \in \mathbb{R}^{2 \times 1}$ by

$$\mathbf{u}_{qc} = \mathbf{J}_f^{-1} \mathbf{u}_f. \tag{8.97}$$

If the equivalent control $\mathbf{u}_f^{eq} \in \mathbb{R}^{2 \times 1}$ is defined as

$$\mathbf{u}_f^{eq} = \mathbf{J}_f \mathbf{A}_n^{-1} \mathbf{p} - \boldsymbol{\Upsilon} + \dot{\mathbf{f}}^{ref} \tag{8.98}$$

dynamics (8.96) becomes

$$\dot{\boldsymbol{\sigma}} = \mathbf{u}_f - \mathbf{u}_f^{eq}. \tag{8.99}$$

The last equation shows that the generalized error dynamics for this task is in the same form as for previously discussed tasks. Therefore, the approach to control synthesis will be analogue to the ones used previously. Assuming that $\boldsymbol{\sigma}$ is available and equivalent control is modeled as $\dot{\mathbf{u}}_f^{eq} = \mathbf{0}$, the equivalent control $\mathbf{u}_f^{eq}$ will be estimated as

$$\begin{aligned} \dot{\mathbf{z}} &= \mathbf{L} (\mathbf{u}_f - \mathbf{z} + \mathbf{L}\boldsymbol{\sigma}) \\ \hat{\mathbf{u}}_f^{eq} &= \mathbf{z} - \mathbf{L}\boldsymbol{\sigma} \end{aligned} \tag{8.100}$$

where $\mathbf{L} \in \mathbb{R}^{2 \times 2}$ is a constant gain matrix given as

$$\mathbf{L} = \mathrm{diag} \, (l_1, \ l_2), \ l_1, l_2 > 0 \tag{8.101}$$

and $\mathbf{u}_f^{eq} + \mathbf{L}\boldsymbol{\sigma} = \mathbf{z} \in \mathbb{R}^{2 \times 1}$ is the intermediate variable in the equivalent control estimation. In this example, the control will be selected to enforce finite-time convergence to the manifold $\boldsymbol{\sigma} = \mathbf{0}$. Therefore, the control vector $\mathbf{u}_f$ is selected as

$$\mathbf{u}_f = \hat{\mathbf{u}}_f^{eq} - \begin{bmatrix} d_1 \, |\sigma_1|^{2\alpha-1} \, \mathrm{sign} \, (\sigma_1) \\ d_2 \, |\sigma_2|^{2\alpha-1} \, \mathrm{sign} \, (\sigma_2) \end{bmatrix}, \ d_1, d_2 > 0, \ 0.5 \leq \alpha < 1. \tag{8.102}$$

When $\mathbf{u}_f$ is calculated, it is necessary to find control forces T_{mc} and T_{sc} as follows

$$\mathbf{T}_c = \begin{bmatrix} T_{mc} \\ T_{sc} \end{bmatrix} = \mathbf{A}_n \mathbf{u}_{qc} = \mathbf{A}_n \mathbf{J}_f^{-1} \mathbf{u}_f. \tag{8.103}$$

Considering (8.83), the input forces on the master and slave side are then given as

$$\begin{bmatrix} T_m \\ T_s \end{bmatrix} = \begin{bmatrix} \hat{T}_{mdis} \\ \hat{T}_{sdis} \end{bmatrix} + \mathbf{A}_n \mathbf{J}_f^{-1} \mathbf{u}_f. \tag{8.104}$$

The controlled system was simulated in MATLAB/Simulink, and the master and slave devices are modeled as 1-DOF manipulators that perform linear motion and have the following parameters

$$a_m(q_m) = a_{mn}[1 + 0.5\sin q_m] \text{ kg}, a_{mn} = 0.1 \text{ kg}$$
$$b_m(q_m, \dot{q}_m) = 15q_m + 0.02\dot{q}_m + 0.05\dot{q}_m \cos q_m \text{ N}$$
$$g_m(q_m) = 9.81q_m^2 \text{ N}$$
$$T_{mext} = 0.1[1 + \cos(4\pi t) + \sin(12\pi t)][H(t - 0.12) - H(t - 0.8)] \text{ N}$$
$$\tag{8.105}$$

$$a_s(q_s) = a_{sn}[1 + 0.5\sin q_s] \text{ kg}, a_{sn} = 0.1 \text{ kg}$$
$$b_s(q_s, \dot{q}_s) = 15q_s + 0.02\dot{q}_s + 0.05\dot{q}_s \cos q_s \text{ N}$$
$$g_s(q_s) = 9.81q_s^2 \text{ N}$$
$$T_{sext} = 0.1[1 + \sin(4\pi t) + \cos(12\pi t)][H(t - 0.12) - H(t - 0.8)] \text{ N}$$
$$\tag{8.106}$$

In (8.105) and (8.106), $H(t)$ is the Heaviside function defined as

$$H(t) = \begin{cases} 0 & t < 0 \\ 1 & t \geq 0. \end{cases} \tag{8.107}$$

The properties of the operator and the environment in the interaction points are

$$\begin{aligned} D_h &= 5 \text{ kg/s} \\ K_h &= 150000 \text{ kg/s}^2 \\ D_e &= 5 \text{ kg/s} \\ K_e &= 225000 \text{ kg/s}^2. \end{aligned} \tag{8.108}$$

The position of the environment is simulated as $q_e(t) = 0.0045\sin(0.5\pi t)$ m. In order to have a realistic illustration of the situations in which the operator would react on the appearance of the interaction force by modifying its motion, operator motion is controlled by a compliance controller. The operator reference motion is $q_h^{ref}(t) = 0.005\sin(2\pi t)$ m. The operator's motion is the result of the compliant control, so its acceleration is

$$\ddot{q}_h = \ddot{q}_h^{ref} + 125\left(\dot{q}_h^{ref} - \dot{q}_h\right) + 2500\left(q_h^{ref} - q_h\right) - 1.85F_h. \tag{8.109}$$

Parameters used above in the control algorithm synthesis are given as follows

$$l_m = l_s = 250, \ c_1 = 100, \ l_1 = l_2 = 2500, \ d_1 = 100, \ d_2 = 250, \ \alpha = 0.8. \tag{8.110}$$

The components of the function space control vector were bounded, to get as realistic simulation results as possible. The absolute value of u_{f1} was limited to 200 m/s², while the limit for the absolute value of u_{f2} was 1000 N/s.

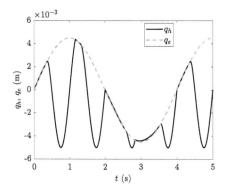

FIGURE 8.20

Operator position and environment position.

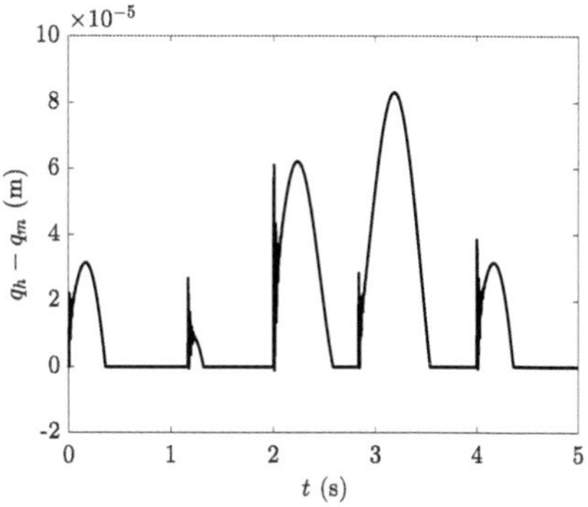

FIGURE 8.21

Difference between the operator and master device positions.

The input forces of the master and slave device, T_m and T_s were also bounded, so that their absolute values cannot exceed 25 N.

Obtained responses are depicted in Figures 8.20–8.28. The diagrams show how environment position affects the operator position, so the compliance of the operator with the environment is obvious and the operator position changes due to the interaction force between the environment and slave device (see Figure 8.20). As shown in Figure 8.21, the error between the operator position and position of the master device is very small, and it illustrates the changes due to the interaction force on the master side. The simulation results illustrate that the interaction between the operator and the remote environment on the slave side is clearly established and that the operator can set both motion and force on the distant environment. The responses from Figures 8.22–8.25 show that both functions in this system are successfully executed. Spikes in the force synchronization tracking error appear when the interaction force on the slave side is being established, but they are compensated by the controller very fast. Out of all control signals, only u_{f2} gets saturated in very short time intervals (see Figures 8.26–8.28).

8.3 Task Involving Elbow Manipulators

In this section, three elbow manipulators will be used to execute an object manipulation task in 3-D space. First, the direct kinematics and dynamic model of an elbow manipulator will be given, and then the control design for the task will be discussed.

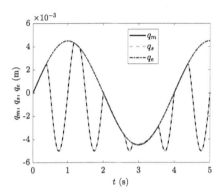

FIGURE 8.22
Master device position, slave device position, and environment position.

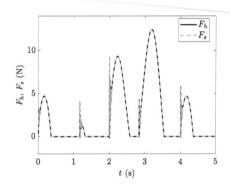

FIGURE 8.23
Master side interaction force and slave side interaction force.

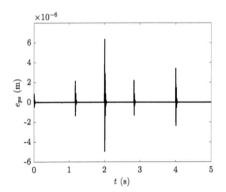

FIGURE 8.24
Tracking error for the position synchronization.

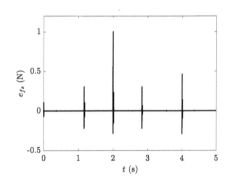

FIGURE 8.25
Tracking error for the force synchronization.

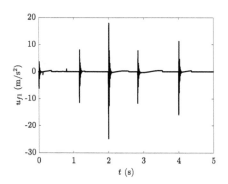

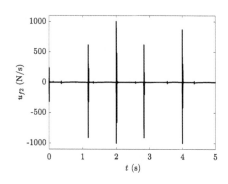

FIGURE 8.26
The first component of the function space control vector.

FIGURE 8.27
The second component of the function space control vector.

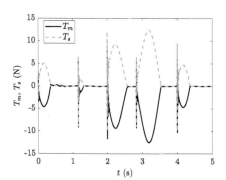

FIGURE 8.28
Input forces on the master side and slave side.

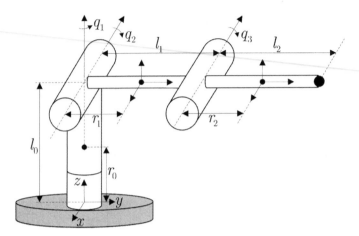

FIGURE 8.29
Elbow manipulator.

8.3.1　Direct Kinematics and Dynamic Model of Elbow Manipulator

An elbow manipulator is depicted in Figure 8.29. In this figure, all joint angles are equal to zero. The direct kinematics and dynamic model of the manipulator are taken from [46]. Parameters for the kinematic model are used as follows. For $i = 0, 1, 2$, l_i denotes the length of link i; r_i is the distance from the previous joint to the center of mass of link i.

The position of the end-effector in the inertial frame attached to the manipulator's base, i.e., in its task space, $\mathbf{x} = [x\ y\ z]^{\mathrm{T}}$ can be expressed in terms of configuration vector $\mathbf{q} = [q_1\ q_2\ q_3]^{\mathrm{T}}$ as

$$\mathbf{x} = \begin{bmatrix} x \\ y \\ z \end{bmatrix} = \begin{bmatrix} -\sin q_1 \left[l_1 \cos q_2 + l_2 \cos \left(q_2 + q_3 \right) \right] \\ \cos q_1 \left[l_1 \cos q_2 + l_2 \cos \left(q_2 + q_3 \right) \right] \\ l_0 - l_1 \sin q_2 - l_2 \sin \left(q_2 + q_3 \right) \end{bmatrix}. \tag{8.111}$$

Taking the first time derivative of the previous equation, the velocity of the end-effector in the task space can be expressed as

$$\dot{\mathbf{x}} = \begin{bmatrix} \dot{x} \\ \dot{y} \\ \dot{z} \end{bmatrix} = \mathbf{J} \begin{bmatrix} \dot{q}_1 \\ \dot{q}_2 \\ \dot{q}_3 \end{bmatrix} = \mathbf{J}\dot{\mathbf{q}} \tag{8.112}$$

where $\mathbf{J} \in \mathbb{R}^{3 \times 3}$ is the manipulator's Jacobian matrix and its elements are given as follows

$$
\begin{aligned}
j_{11} &= -\cos q_1 \left[l_1 \cos q_2 + l_2 \cos (q_2 + q_3) \right] \\
j_{12} &= \sin q_1 \left[l_1 \sin q_2 + l_2 \sin (q_2 + q_3) \right] \\
j_{13} &= \sin q_1 l_2 \sin (q_2 + q_3) \\
j_{21} &= -\sin q_1 \left[l_1 \cos q_2 + l_2 \cos (q_2 + q_3) \right] \\
j_{22} &= -\cos q_1 \left[l_1 \sin q_2 + l_2 \sin (q_2 + q_3) \right] \\
j_{23} &= -\cos q_1 l_2 \sin (q_2 + q_3) \\
j_{31} &= 0 \\
j_{32} &= -l_1 \cos q_2 - l_2 \cos (q_2 + q_3) \\
j_{33} &= -l_2 \cos (q_2 + q_3) .
\end{aligned}
\tag{8.113}
$$

In the derivation of the dynamic model, to each link a frame is attached at the center of mass and aligned with the principle inertia axes of the link. In Figure 8.29, these frames are displaced with respect to the inertial frame attached to the base, i.e., no rotation is performed. For the dynamic model, m_i is the mass of the link i ($i = 0, 1, 2$), while I_{xi}, I_{yi}, and I_{zi} are the moments of inertia about the x-, y-, and z-axes of the ith link frame. The dynamic model in the configuration space can be written in the following form

$$
\left.
\begin{aligned}
\dot{\mathbf{q}} &= \mathbf{v} \\
\mathbf{A}(\mathbf{q})\dot{\mathbf{v}} + \mathbf{b}(\mathbf{q}, \mathbf{v}) + \mathbf{g}(\mathbf{q}) &= \mathbf{T}
\end{aligned}
\right\}
\tag{8.114}
$$

where $\mathbf{q} = [q_1 \; q_2 \; q_3]^{\mathrm{T}}$ denotes the configuration vector of the joint angles, $\mathbf{v} = [v_1 \; v_2 \; v_3]^{\mathrm{T}}$ is the configuration space velocity vector, $\mathbf{A}(\mathbf{q}) \in \mathbb{R}^{3 \times 3}$ stands for the symmetric positive definite kinetic energy matrix (inertia matrix), $\mathbf{b}(\mathbf{q}, \mathbf{v}) \in \mathbb{R}^{3 \times 1}$ stands for the vector of Coriolis forces and centripetal forces, $\mathbf{g}(\mathbf{q}) \in \mathbb{R}^{3 \times 1}$ is the vector of gravity terms, $\mathbf{T} \in \mathbb{R}^{3 \times 1}$ is the vector of joint torques (sometimes it will be referred to $\mathbf{T}$ as the input force vector in the configuration space).

For a shorter version, the following notation will be used in the further derivation of the dynamic model

$$
s_i = \sin q_i, \; c_i = \cos q_i, s_{ij} = \sin (q_i + q_j), c_{ij} = \sin (q_i + q_j), \; i, j = 1, 2, 3.
\tag{8.115}
$$

The elements of the kinetic energy matrix are then given as

$$
\begin{aligned}
a_{11} &= I_{y2} s_2^2 + I_{y3} s_{23}^2 + I_{z1} + I_{z2} c_2^2 + I_{z3} c_{23}^2 + m_2 r_1^2 c_2^2 + m_3 \left(l_1 c_2 + r_2 c_{23} \right)^2 \\
a_{12} &= a_{21} = 0 \\
a_{13} &= a_{31} = 0 \\
a_{22} &= I_{x2} + I_{x3} + m_3 l_1^2 + m_2 r_1^2 + m_3 r_2^2 + 2 m_3 l_1 r_2 c_3 \\
a_{23} &= a_{32} = I_{x3} + m_3 r_2^2 + m_3 l_1 r_2 c_3 \\
a_{33} &= I_{x3} + m_3 r_2^2 .
\end{aligned}
\tag{8.116}
$$

Vector $\mathbf{b}(\mathbf{q}, \mathbf{v})$ can be expressed in the form

$$
\mathbf{b}(\mathbf{q}, \mathbf{v}) = \mathbf{K}(\mathbf{q}, \mathbf{v})\mathbf{v}.
\tag{8.117}
$$

The elements of the matrix $\mathbf{K}(\mathbf{q}, \mathbf{v}) \in \mathbb{R}^{3 \times 3}$ appearing in (8.117) can be calculated as

$$k_{ij}(\mathbf{q}, \mathbf{v}) = \sum_{k=1}^{3} \Gamma_{ijk} v_k \tag{8.118}$$

where the nonzero Γ_{ijk} values are

$$
\begin{aligned}
\Gamma_{112} &= \left(I_{y2} - I_{z2} - m_2 r_1^2\right) c_2 s_2 + (I_{y3} - I_{z3}) c_{23} s_{23} \\
&\quad - m_3 (l_1 c_2 + r_2 c_{23}) (l_1 s_2 + r_2 s_{23}) \\
\Gamma_{113} &= (I_{y3} - I_{z3}) c_{23} s_{23} - m_3 r_2 s_{23} (l_1 c_2 + r_2 c_{23}) \\
\Gamma_{121} &= \left(I_{y2} - I_{z2} - m_2 r_1^2\right) c_2 s_2 + (I_{y3} - I_{z3}) c_{23} s_{23} \\
&\quad - m_3 (l_1 c_2 + r_2 c_{23}) (l_1 s_2 + r_2 s_{23}) \\
\Gamma_{131} &= (I_{y3} - I_{z3}) c_{23} s_{23} - m_3 r_2 s_{23} (l_1 c_2 + r_2 c_{23})
\end{aligned}
\tag{8.119}
$$

$$
\begin{aligned}
\Gamma_{211} &= \left(I_{z2} - I_{y2} + m_2 r_1^2\right) c_2 s_2 + (I_{z3} - I_{y3}) c_{23} s_{23} \\
&\quad + m_3 (l_1 c_2 + r_2 c_{23}) (l_1 s_2 + r_2 s_{23}) \\
\Gamma_{223} &= - l_1 m_3 r_2 s_3 \\
\Gamma_{232} &= - l_1 m_3 r_2 s_3 \\
\Gamma_{233} &= - l_1 m_3 r_2 s_3 \\
\Gamma_{311} &= (I_{z3} - I_{y3}) c_{23} s_{23} + m_3 r_2 s_{23} (l_1 c_2 + r_2 c_{23}) \\
\Gamma_{322} &= l_1 m_3 r_2 s_3 \quad .
\end{aligned}
\tag{8.120}
$$

Finally, the elements of the vector $\mathbf{g}(\mathbf{q})$ can be expressed as follows

$$
\mathbf{g}(\mathbf{q}) = \begin{bmatrix} 0 \\ - (m_2 g r_1 + m_3 g l_1) \cos q_2 - m_3 g r_2 \cos (q_2 + q_3) \\ - m_3 g r_2 \cos (q_2 + q_3) \end{bmatrix}
\tag{8.121}
$$

where g is the magnitude of the gravitational acceleration which is directed along the negative z-axis in Figure 8.29.

The presented dynamic model was used in MATLAB/Simulink with the parameters given in Table 8.2.

8.3.2   Object Manipulation Task in 3-D Space

In the execution of the object manipulation task, three identical manipulators are used, and their parameters are given in Table 8.2. The manipulators are positioned in such a way that the x- and y-axes of the inertial frames attached to their bases lie in a plane, and z-axes are all parallel to each other. The manipulators are positioned on a circle with radius P and rotated for $2\pi/3$ with respect to each other. Looking from above, their displacement can be represented as in Figure 8.30. Superscripts (1), (2), and (3) refer to the frames attached to the first, second and third manipulator, respectively. For this

TABLE 8.2

Parameters of elbow
manipulator.

Parameter	Value
l_0	1 m
r_0	0.5 m
l_1	1 m
r_1	0.5 m
l_2	1 m
r_2	0.5 m
m_1	2 kg
m_2	1.6 kg
m_3	1.7 kg
I_{x1}	0.1707 kgm^2
I_{y1}	0.1707 kgm^2
I_{z1}	0.0081 kgm^2
I_{x2}	0.1359 kgm^2
I_{y2}	0.0051 kgm^2
I_{z2}	0.1359 kgm^2
I_{x3}	0.1444 kgm^2
I_{y3}	0.0054 kgm^2
I_{z3}	0.1444 kgm^2

system, the global coordinate frame is an inertial frame placed in the center
of the circle on which manipulators are positioned. All z-axes in this system
have the same direction, coming out of the plane of the paper. Superscript
(G) refers to the global coordinate frame.

Task space coordinates of the manipulators can be expressed in the global
coordinate frame using three homogeneous transformations. Let us adopt a
notation in which $\nu_i^{(i)}$, $\nu = x$, y, z, $i = 1$, 2, 3, represents the ν coordinate of
the ith manipulator expressed in its local inertial coordinate frame (local task
space), while $\nu_i^{(G)}$ is the same coordinate expressed in the global coordinate
frame. The homogeneous transformations can be expressed as

$$\begin{bmatrix} x_i^{(G)} \\ y_i^{(G)} \\ z_i^{(G)} \\ 1 \end{bmatrix} = \underbrace{\begin{bmatrix} \mathbf{R}_i & \mathbf{d}_i \\ \mathbf{0}_{1\times3} & 1 \end{bmatrix}}_{\mathbf{H}_i} \begin{bmatrix} x_i^{(i)} \\ y_i^{(i)} \\ z_i^{(i)} \\ 1 \end{bmatrix} , \; i = 1, 2, 3. \tag{8.122}$$

For each manipulator, the transformation matrix $\mathbf{H}_i \in \mathbb{R}^{4\times4}$ is different. It
contains the orthogonal matrix $\mathbf{R}_i \in \mathbb{R}^{3\times3}$ and vector $\mathbf{d}_i \in \mathbb{R}^{3\times1}$, which are
for the three manipulators given as follows

$$\mathbf{R}_1 = \begin{bmatrix} -1/2 & \sqrt{3}/2 & 0 \\ -\sqrt{3}/2 & -1/2 & 0 \\ 0 & 0 & 1 \end{bmatrix}, \; \mathbf{d}_1 = \begin{bmatrix} -P\sqrt{3}/2 \\ P/2 \\ 0 \end{bmatrix} \tag{8.123}$$

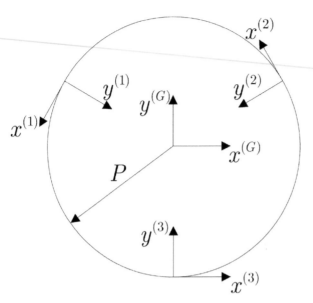

FIGURE 8.30
Displacement of three elbow manipulators.

$$\mathbf{R}_2 = \begin{bmatrix} -1/2 & -\sqrt{3}/2 & 0 \\ \sqrt{3}/2 & -1/2 & 0 \\ 0 & 0 & 1 \end{bmatrix}, \ \mathbf{d}_2 = \begin{bmatrix} P\sqrt{3}/2 \\ P/2 \\ 0 \end{bmatrix} \quad (8.124)$$

$$\mathbf{R}_3 = \begin{bmatrix} 1 & 0 & 0 \\ 0 & 1 & 0 \\ 0 & 0 & 1 \end{bmatrix}, \ \mathbf{d}_3 = \begin{bmatrix} 0 \\ -P \\ 0 \end{bmatrix} \quad (8.125)$$

Dynamics of the three manipulators in the object manipulation task can be written as

$$\left. \begin{aligned} \dot{\mathbf{q}}_1 &= \mathbf{v}_1 \\ \mathbf{A}_1(\mathbf{q}_1)\dot{\mathbf{v}}_1 + \mathbf{b}_1(\mathbf{q}_1, \mathbf{v}_1) + \mathbf{g}_1(\mathbf{q}_1) &= \mathbf{T}_1 + \mathbf{T}_1^r \end{aligned} \right\} \quad (8.126)$$

$$\left. \begin{aligned} \dot{\mathbf{q}}_2 &= \mathbf{v}_2 \\ \mathbf{A}_2(\mathbf{q}_2)\dot{\mathbf{v}}_2 + \mathbf{b}_2(\mathbf{q}_2, \mathbf{v}_2) + \mathbf{g}_2(\mathbf{q}_2) &= \mathbf{T}_2 + \mathbf{T}_2^r \end{aligned} \right\} \quad (8.127)$$

$$\left. \begin{aligned} \dot{\mathbf{q}}_3 &= \mathbf{v}_3 \\ \mathbf{A}_3(\mathbf{q}_3)\dot{\mathbf{v}}_3 + \mathbf{b}_3(\mathbf{q}_3, \mathbf{v}_3) + \mathbf{g}_3(\mathbf{q}_3) &= \mathbf{T}_3 + \mathbf{T}_3^r \end{aligned} \right\} \quad (8.128)$$

where subscripts 1, 2 and 3 refer to the first, second and third manipulator. $\mathbf{T}_k^r \in \mathbb{R}^{3 \times 1}, k = 1, 2, 3$ stands for the vector of joint torques that appear due to the interaction between the kth manipulator and a grasped object. The components of the input force vectors in the configuration space, $\mathbf{T}_1$, $\mathbf{T}_2$ and $\mathbf{T}_3$, are given as

$$\mathbf{T}_1 = \begin{bmatrix} T_{11} \\ T_{12} \\ T_{13} \end{bmatrix}, \ \mathbf{T}_2 = \begin{bmatrix} T_{21} \\ T_{22} \\ T_{23} \end{bmatrix}, \ \mathbf{T}_3 = \begin{bmatrix} T_{31} \\ T_{32} \\ T_{33} \end{bmatrix}. \tag{8.129}$$

The dynamics of all manipulators can be written in the same form. For the ith manipulator, $i = 1, 2, 3$, it will be

$$\left. \begin{aligned} \dot{\mathbf{q}}_i &= \mathbf{v}_i \\ \mathbf{A}_i(\mathbf{q}_i)\dot{\mathbf{v}}_i &= \mathbf{T}_i - \underbrace{[\mathbf{b}_i(\mathbf{q}_i, \mathbf{v}_i) + \mathbf{g}_i(\mathbf{q}_i) - \mathbf{T}_i^r]}_{\mathbf{T}_{idis}} \end{aligned} \right\} \tag{8.130}$$

where $\mathbf{T}_{idis} \in \mathbb{R}^{3 \times 1}$ is the disturbance for the ith manipulator. For each manipulator, the acting disturbance is estimated assuming that $\mathbf{v}_i$ is available and with the disturbance modeled as $\dot{\mathbf{T}}_{idis} = \mathbf{0}$. The vector $\mathbf{T}_{idis}$ is estimated as follows

$$\left. \begin{aligned} \dot{\mathbf{z}}_i &= \mathbf{L}_i \mathbf{A}_i^{-1} \left(\mathbf{T}_i - \mathbf{z}_i + \mathbf{L}_i \mathbf{v}_i \right) \\ \hat{\mathbf{T}}_{idis} &= \mathbf{z}_i - \mathbf{L}_i \mathbf{v}_i \end{aligned} \right. \tag{8.131}$$

where $\mathbf{L}_i \in \mathbb{R}^{3 \times 3}$ is a constant gain matrix given as

$$\mathbf{L}_i = \text{diag}\left(l_{i1},\ l_{i2},\ l_{i3} \right),\ l_{ij} > 0,\ j = 1, 2, 3. \tag{8.132}$$

and $\mathbf{T}_{idis} + \mathbf{L}_i \mathbf{v}_i = \mathbf{z}_i \in \mathbb{R}^{3 \times 1}$ is the intermediate variable in the disturbance estimation. It will be assumed that the input force vector $\mathbf{T}_i \in \mathbb{R}^{3 \times 1}$ consists of the estimated disturbance $\hat{\mathbf{T}}_{idis}$ and additional term $\mathbf{T}_{ic} \in \mathbb{R}^{3 \times 1}$ which is to be calculated by a high-level controller. Therefore, it can be written

$$\mathbf{T}_i = \hat{\mathbf{T}}_{idis} + \mathbf{T}_{ic}. \tag{8.133}$$

Now, dynamics (8.130) becomes

$$\left. \begin{aligned} \dot{\mathbf{q}}_i &= \mathbf{v}_i \\ \mathbf{A}_i(\mathbf{q}_i)\dot{\mathbf{v}}_i &= \mathbf{T}_{ic} - \underbrace{\left(\mathbf{T}_{idis} - \hat{\mathbf{T}}_{idis} \right)}_{\mathbf{p}_i}. \end{aligned} \right\} \tag{8.134}$$

In (8.134), $\mathbf{p}_i \in \mathbb{R}^{3 \times 1}$ is the disturbance estimation error appearing in the estimation of the disturbance for the ith manipulator. In a compact form, the dynamics of the system that consists of the three manipulators with disturbance observes can be written as

$$\left. \begin{aligned} \dot{\tilde{\mathbf{q}}} &= \tilde{\mathbf{v}} \\ \tilde{\mathbf{A}}(\tilde{\mathbf{q}})\dot{\tilde{\mathbf{v}}} &= \tilde{\mathbf{T}}_c - \tilde{\mathbf{p}}. \end{aligned} \right\} \tag{8.135}$$

The matrices and vectors that appear in (8.135) are given as

$$\tilde{\mathbf{A}} = \begin{bmatrix} \mathbf{A}_1 & \mathbf{0}^{3\times3} & \mathbf{0}^{3\times3} \\ \mathbf{0}^{3\times3} & \mathbf{A}_2 & \mathbf{0}^{3\times3} \\ \mathbf{0}^{3\times3} & \mathbf{0}^{3\times3} & \mathbf{A}_3 \end{bmatrix}, \ \tilde{\mathbf{q}} = \begin{bmatrix} \mathbf{q}_1 \\ \mathbf{q}_2 \\ \mathbf{q}_3 \end{bmatrix}, \ \tilde{\mathbf{v}} = \begin{bmatrix} \mathbf{v}_1 \\ \mathbf{v}_2 \\ \mathbf{v}_3 \end{bmatrix},$$

$$\tilde{\mathbf{T}}_c = \begin{bmatrix} \mathbf{T}_{1c} \\ \mathbf{T}_{2c} \\ \mathbf{T}_{3c} \end{bmatrix}, \ \tilde{\mathbf{p}} = \begin{bmatrix} \mathbf{p}_1 \\ \mathbf{p}_2 \\ \mathbf{p}_3 \end{bmatrix}. \tag{8.136}$$

Since $\tilde{\mathbf{A}} \in \mathbb{R}^{9\times9}$ is a nonsingular matrix, dynamics (8.135) can also be written in the following form

$$\left. \begin{array}{l} \dot{\tilde{\mathbf{q}}} = \tilde{\mathbf{v}} \\ \dot{\tilde{\mathbf{v}}} = \tilde{\mathbf{u}}_{\tilde{q}c} - \tilde{\mathbf{A}}^{-1}\tilde{\mathbf{p}}, \ \tilde{\mathbf{u}}_{\tilde{q}c} = \tilde{\mathbf{A}}^{-1}\tilde{\mathbf{T}}_c. \end{array} \right\} \tag{8.137}$$

During the execution of the object manipulation task, one would like to have end-effectors of the three manipulators forming an equilateral triangle, which will lie in a plane parallel to the x-y plane of the global coordinate frame. At the same time, the position of the triangle's center (center of circumscribed circle) should be controlled, together with the orientation of the triangle. The radius of the circumscribed circle will control the grasping force applied to the manipulated object. For simplicity, the manipulated object is assumed to be a ball-shaped object. Looking in a plane parallel to the x-y plane, this situation can be illustrated as in Figure 8.31. In the figure, positions of the end-effectors are given as

$$\mathbf{P}_1 = \begin{bmatrix} x_1^{(G)} \\ y_1^{(G)} \\ z_1^{(G)} \end{bmatrix}, \ \mathbf{P}_2 = \begin{bmatrix} x_2^{(G)} \\ y_2^{(G)} \\ z_2^{(G)} \end{bmatrix}, \ \mathbf{P}_3 = \begin{bmatrix} x_3^{(G)} \\ y_3^{(G)} \\ z_3^{(G)} \end{bmatrix} \tag{8.138}$$

while the center of the triangle is

$$\mathbf{P}_C = \begin{bmatrix} x_C^{(G)} \\ y_C^{(G)} \\ z_C^{(G)} \end{bmatrix}. \tag{8.139}$$

Distance R stands for the radius of the circumscribed circle, while d is the side length of the triangle. The angle θ describes the orientation of the triangle, and it is the angle of rotation around the axis which is passing through the center and has the same direction as the z-axis of the global coordinate frame, where rotation is in the positive (counter-clockwise) direction.

For the object grasped by the manipulators, the grasping force will be modeled as

$$F_g = K_e (2R_0 - 2R) - D_e 2\dot{R}. \tag{8.140}$$

In (8.140), K_e and D_e are stiffness and damping coefficients characterizing the manipulated object in the contact points between the manipulators and

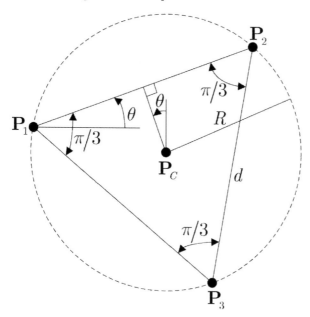

FIGURE 8.31
Three manipulators forming an equilateral triangle.

object, R is the radius of circumscribed circle defined by the positions of the three manipulators ($2R$ is diameter), while R_0 is a known constant. This constant is determined from the condition that the grasping force exists for $2R < 2R_0$.

In the design of the control algorithm, the following approach is adopted. The manipulators are controlled to form an equilateral triangle in a plane parallel to the x-y plane. The reference side length is determined from desired grasping force dynamics. If the reference grasping force is $F_g^{ref}(t)$, and it is a differentiable function of time, the tracking error for the grasping force is

$$e_F = F_g - F_g^{ref}. \tag{8.141}$$

If the desired dynamics of the grasping force error is expressed by

$$\dot{e}_F = -k_F e_F, k_F > 0 \tag{8.142}$$

where k_F is known positive constant, one can determine the desired second-order dynamics of the circumscribed circle's radius $\ddot{R}^{ref}$ that will enforce the desired dynamics of the grasping force error as

$$\ddot{R}^{ref} = \frac{k_F}{2D_e} e_F - \frac{K_e}{D_e} \dot{R}^{ref}. \tag{8.143}$$

Therefore, $u_v \triangleq \ddot{R}^{ref}$ can be considered as the virtual control input that will enforce the desired dynamics of the grasping force tracking error. The dynamics (8.143) can be modeled, and its output is the desired radius of circumscribed circle R^{ref}. When R^{ref} is known, the reference side length is

$$d^{ref} = \frac{3R^{ref}}{\sqrt{3}}. \tag{8.144}$$

If positions of the manipulators' end-effectors in the global coordinate frame are known, one can determine the center and radius of the circumscribed circle for the triangle whose vertices are positions of the end-effectors. If the positions are represented as in (8.138), the radius of the circumscribed circle is

$$R = \frac{\|\mathbf{P}_1 - \mathbf{P}_2\| \, \|\mathbf{P}_2 - \mathbf{P}_3\| \, \|\mathbf{P}_3 - \mathbf{P}_1\|}{2 \, \|(\mathbf{P}_1 - \mathbf{P}_2) \times (\mathbf{P}_2 - \mathbf{P}_3)\|}. \tag{8.145}$$

The coordinates of the center can be obtained as

$$\mathbf{P}_C = \begin{bmatrix} x_C^{(G)} \\ y_C^{(G)} \\ z_C^{(G)} \end{bmatrix} = \alpha \mathbf{P}_1 + \beta \mathbf{P}_2 + \gamma \mathbf{P}_3 \tag{8.146}$$

where

$$\begin{aligned} \alpha &= \frac{|\mathbf{P}_2 - \mathbf{P}_3|^2 (\mathbf{P}_1 - \mathbf{P}_2) \cdot (\mathbf{P}_1 - \mathbf{P}_3)}{2 |(\mathbf{P}_1 - \mathbf{P}_2) \times (\mathbf{P}_2 - \mathbf{P}_3)|^2} \\ \beta &= \frac{|\mathbf{P}_1 - \mathbf{P}_3|^2 (\mathbf{P}_2 - \mathbf{P}_1) \cdot (\mathbf{P}_2 - \mathbf{P}_3)}{2 |(\mathbf{P}_1 - \mathbf{P}_2) \times (\mathbf{P}_2 - \mathbf{P}_3)|^2} \\ \gamma &= \frac{|\mathbf{P}_1 - \mathbf{P}_2|^2 (\mathbf{P}_3 - \mathbf{P}_1) \cdot (\mathbf{P}_3 - \mathbf{P}_2)}{2 |(\mathbf{P}_1 - \mathbf{P}_2) \times (\mathbf{P}_2 - \mathbf{P}_3)|^2}. \end{aligned} \tag{8.147}$$

Due to the grasping force, reaction forces $\mathbf{F}_i^r \in \mathbb{R}^{3 \times 1}$, $i = 1, 2, 3$, appear to be acting on the manipulators. These forces, expressed in the local frames attached to the bases of the manipulators, are modeled as

$$\mathbf{F}_i^r = F_g \mathbf{R}_i^{-1} \frac{\mathbf{P}_i - \mathbf{P}_C}{|\mathbf{P}_i - \mathbf{P}_C|}. \tag{8.148}$$

Now, $\mathbf{T}_i^r, i = 1, 2, 3$ from (8.126)–(8.128) are

$$\mathbf{T}_i^r = \mathbf{J}_i^T \mathbf{F}_i^r \tag{8.149}$$

where $\mathbf{J}_i \in \mathbb{R}^{3 \times 3}$, $i = 1, 2, 3$, are the Jacobian matrices of the manipulators.

In the design of the control algorithm, it is assumed that the reference side length of the equilateral triangle that should be formed by the end-effectors is calculated from (8.144). Additionally, $\mathbf{P}_C^{ref}(t) = \left[x_C^{ref}(t) \ \ y_C^{ref}(t) \ \ z_C^{ref}(t) \right]^T$ is given as the reference for the center of the triangle, and it is two times differentiable vector expressed in the global coordinate frame. Finally, the reference orientation of the triangle is two times differentiable function $\theta^{ref}(t)$.

Nine functions to be controlled are specified as follows

$$\varphi\left(\mathbf{q}_1, \mathbf{q}_2, \mathbf{q}_3\right) = \begin{bmatrix} \varphi_1\left(\mathbf{q}_1, \mathbf{q}_2\right) \\ \varphi_2\left(\mathbf{q}_1, \mathbf{q}_2\right) \\ \varphi_3\left(\mathbf{q}_1, \mathbf{q}_2\right) \\ \varphi_4\left(\mathbf{q}_3\right) \\ \varphi_5\left(\mathbf{q}_3\right) \\ \varphi_6\left(\mathbf{q}_3\right) \\ \varphi_7\left(\mathbf{q}_1, \mathbf{q}_2\right) \\ \varphi_8\left(\mathbf{q}_1, \mathbf{q}_2\right) \\ \varphi_9\left(\mathbf{q}_1, \mathbf{q}_3\right) \end{bmatrix} = \begin{bmatrix} x_1^{(G)} - x_2^{(G)} \\ y_1^{(G)} - y_2^{(G)} \\ z_1^{(G)} - z_2^{(G)} \\ x_3^{(G)} \\ y_3^{(G)} \\ z_3^{(G)} \\ x_1^{(G)} + x_2^{(G)} \\ y_1^{(G)} + y_2^{(G)} \\ z_1^{(G)} - z_3^{(G)} \end{bmatrix} \tag{8.150}$$

and their references are

$$\varphi^{ref}\left(t\right) = \begin{bmatrix} \varphi_1^{ref}\left(t\right) \\ \varphi_2^{ref}\left(t\right) \\ \varphi_3^{ref}\left(t\right) \\ \varphi_4^{ref}\left(t\right) \\ \varphi_5^{ref}\left(t\right) \\ \varphi_6^{ref}\left(t\right) \\ \varphi_7^{ref}\left(t\right) \\ \varphi_8^{ref}\left(t\right) \\ \varphi_9^{ref}\left(t\right) \end{bmatrix} = \begin{bmatrix} -d^{ref}\cos\theta^{ref} \\ -d^{ref}\sin\theta^{ref} \\ 0 \\ x_3^{ref} \\ y_3^{ref} \\ z_3^{ref} \\ 2x_C^{ref} - \frac{d^{ref}\sqrt{3}}{3}\sin\theta^{ref} \\ 2y_C^{ref} + \frac{d^{ref}\sqrt{3}}{3}\cos\theta^{ref} \\ 0 \end{bmatrix}. \tag{8.151}$$

In (8.151), x_3^{ref}, y_3^{ref}, and z_3^{ref} are calculated as

$$\begin{bmatrix} x_3^{ref} \\ y_3^{ref} \\ z_3^{ref} \end{bmatrix} = \begin{bmatrix} \cos\theta^{ref} & -\sin\theta^{ref} & 0 \\ \sin\theta^{ref} & \theta^{ref} & 0 \\ 0 & 0 & 1 \end{bmatrix} \begin{bmatrix} 0 \\ -\frac{d^{ref}\sqrt{3}}{3} \\ 0 \end{bmatrix} + \mathbf{P}_C^{ref}. \tag{8.152}$$

If $\varphi = \varphi^{ref}$ is enforced, the manipulators' end-effectors will form the desired equilateral triangle and the triangle will lie in a plane parallel to the x-y plane, with side length d^{ref}, center in $\mathbf{P}_C^{ref}$, and it will be rotated for θ^{ref} around the axis which passes through the center and has the same direction as the z-axis of the global coordinate frame, where the rotation is in the positive direction (like in Figure 8.31).

The defined functions φ_i, $i = 1, 2, \ldots, 9$ depend on the configuration vectors only. Thus, the function vector is

$$\mathbf{f} = \dot{\varphi} \tag{8.153}$$

The function vector $\mathbf{f}$ can be further expressed as

$$\dot{\mathbf{f}} = \mathbf{J}_f \tilde{\mathbf{v}} \tag{8.154}$$

where $\mathbf{J}_f \in \mathbb{R}^{9 \times 9}$ is the function Jacobian matrix defined as

$$\mathbf{J}_f = \begin{bmatrix} \mathbf{R}_1 \mathbf{J}_1 & -\mathbf{R}_2 \mathbf{J}_2 & \mathbf{0}_{3 \times 3} \\ \mathbf{0}_{3 \times 3} & \mathbf{0}_{3 \times 3} & \mathbf{R}_3 \mathbf{J}_3 \\ \mathbf{R}_1 \mathbf{J}_1 & \mathbf{W}_1 & \mathbf{W}_2 \end{bmatrix}. \tag{8.155}$$

In (8.155), the matrices $\mathbf{W}_1 \in \mathbb{R}^{3 \times 3}$ and $\mathbf{W}_2 \in \mathbb{R}^{3 \times 3}$ are calculated as follows

$$\begin{aligned} \mathbf{W}_1 &= \mathbf{S}_1 \mathbf{R}_2 \mathbf{J}_2, \ \mathbf{S}_1 = \text{diag}\,(1,\ 1,\ 0) \\ \mathbf{W}_2 &= \mathbf{S}_2 \mathbf{R}_3 \mathbf{J}_3, \ \mathbf{S}_2 = \text{diag}\,(0,\ 0,\ -1). \end{aligned} \tag{8.156}$$

The reference for the function vector is

$$\mathbf{f}^{ref} = \dot{\boldsymbol{\varphi}}^{ref} - \mathbf{C}\left(\boldsymbol{\varphi} - \boldsymbol{\varphi}^{ref}\right) \tag{8.157}$$

where $\mathbf{C} \in \mathbb{R}^{9 \times 9}$ is a constant diagonal matrix with positive diagonal entries

$$\mathbf{C} = \text{diag}\,(c_1, c_2, \ldots, c_9), \ c_i > 0, \ i = 1, 2, \ldots, 9. \tag{8.158}$$

It has to be noted that constants c_i from (8.158) do not have the meaning given in (8.115) since that notation was used only in the derivation of the dynamic model.

In the controlled system, the tracking error vector $\mathbf{e} \in \mathbb{R}^{9 \times 1}$ is

$$\mathbf{e} = \begin{bmatrix} e_1 & e_2 & \cdots & e_9 \end{bmatrix}^{\mathrm{T}} = \boldsymbol{\varphi} - \boldsymbol{\varphi}^{ref}. \tag{8.159}$$

The generalized error is defined by

$$\boldsymbol{\sigma} = \begin{bmatrix} \sigma_1 & \sigma_2 & \cdots & \sigma_9 \end{bmatrix}^{\mathrm{T}} = \mathbf{f} - \mathbf{f}^{ref}. \tag{8.160}$$

The first-order dynamics of the generalized error is

$$\dot{\boldsymbol{\sigma}} = \mathbf{J}_f \dot{\tilde{\mathbf{v}}} + \underbrace{\dot{\mathbf{J}}_f \tilde{\mathbf{v}}}_{\boldsymbol{\Upsilon}} - \dot{\mathbf{f}}^{ref}. \tag{8.161}$$

Considering (8.137), it further becomes

$$\dot{\boldsymbol{\sigma}} = \mathbf{J}_f \tilde{\mathbf{u}}_{\tilde{q}c} - \left(\mathbf{J}_f \tilde{\mathbf{A}}^{-1} \tilde{\mathbf{p}} - \boldsymbol{\Upsilon} + \dot{\mathbf{f}}^{ref}\right). \tag{8.162}$$

Let us now introduce the control vector in the function space $\begin{bmatrix} u_{f1} & \cdots & u_{f9} \end{bmatrix}^{\mathrm{T}} = \mathbf{u}_f \in \mathbb{R}^{9 \times 1}$ which is related to the control acceleration vector $\tilde{\mathbf{u}}_{\tilde{q}c} \in \mathbb{R}^{9 \times 1}$ by

$$\tilde{\mathbf{u}}_{\tilde{q}c} = \mathbf{J}_f^{-1} \mathbf{u}_f. \tag{8.163}$$

Thus, it is assumed that $\mathbf{J}_f$ stays a nonsingular matrix for the entire time of operation. If the equivalent control is defined as

$$\mathbf{u}_f^{eq} = \mathbf{J}_f \tilde{\mathbf{A}}^{-1} \tilde{\mathbf{p}} - \boldsymbol{\Upsilon} + \dot{\mathbf{f}}^{ref} \tag{8.164}$$

dynamics (8.162) can be written as

$$\dot{\boldsymbol{\sigma}} = \mathbf{u}_f - \mathbf{u}_f^{eq}. \tag{8.165}$$

The first-order dynamics of the generalized error (8.165) has the same form as in the previously discussed examples. Assuming that $\boldsymbol{\sigma}$ is available and with the equivalent control modeled as $\dot{\mathbf{u}}_f^{eq} = \mathbf{0}$, the equivalent control can be estimated as

$$\begin{aligned} \dot{\mathbf{z}} &= \mathbf{L}\left(\mathbf{u}_f - \mathbf{z} + \mathbf{L}\boldsymbol{\sigma}\right) \\ \hat{\mathbf{u}}_f^{eq} &= \mathbf{z} - \mathbf{L}\boldsymbol{\sigma} \end{aligned} \tag{8.166}$$

where $\mathbf{L} \in \mathbb{R}^{9 \times 9}$ is a constant gain matrix given as

$$\mathbf{L} = \operatorname{diag}\left(l_1, l_2, \ldots, l_9\right), \ l_i > 0, \ i = 1, 2, \ldots, 9. \tag{8.167}$$

and $\mathbf{u}_f^{eq} + \mathbf{L}\boldsymbol{\sigma} = \mathbf{z} \in \mathbb{R}^{9 \times 1}$ is the intermediate variable used in the equivalent control estimation. The function space control $\mathbf{u}_f$ is selected to enforce exponential convergence of the generalized error components, and it is calculated as

$$\mathbf{u}_f = \mathbf{u}_f^{eq} - \mathbf{D}\boldsymbol{\sigma} \tag{8.168}$$

where $\mathbf{D} \in \mathbb{R}^{9 \times 9}$ is a constant diagonal matrix

$$\mathbf{D} = \operatorname{diag}\left(d_1, d_2, \ldots, d_9\right), \ d_i > 0, \ i = 1, 2, \ldots, 9. \tag{8.169}$$

When $\mathbf{u}_f$ is calculated, it is necessary to map it back to the configuration space, so appropriate torques can be applied to the joints of the manipulators. Considering (8.137) and (8.163), the control input in the configuration space that comes from the high-level controller, $\tilde{\mathbf{T}}_c$, is

$$\tilde{\mathbf{T}}_c = \begin{bmatrix} \mathbf{T}_{1c} \\ \mathbf{T}_{2c} \\ \mathbf{T}_{3c} \end{bmatrix} = \tilde{\mathbf{A}} \mathbf{J}_f^{-1} \mathbf{u}_f. \tag{8.170}$$

The high-level controller calculates $\tilde{\mathbf{T}}_c$ based on the controlled functions. Taking (8.133) into account, the total input forces applied to the manipulators can be expressed

$$\begin{bmatrix} \mathbf{T}_1 \\ \mathbf{T}_2 \\ \mathbf{T}_3 \end{bmatrix} = \begin{bmatrix} \hat{\mathbf{T}}_{1dis} \\ \hat{\mathbf{T}}_{2dis} \\ \hat{\mathbf{T}}_{3dis} \end{bmatrix} + \tilde{\mathbf{A}} \mathbf{J}_f^{-1} \mathbf{u}_f. \tag{8.171}$$

In (8.170) it is necessary to calculate inversion of the matrix $\mathbf{J}_f$, which is a matrix of order nine. In order to speed-up this calculation, the inverse matrix $\mathbf{J}_f^{-1}$ can be calculated using the following procedure

$$\mathbf{J}_f = \begin{bmatrix} \underbrace{\mathbf{R}_1 \mathbf{J}_1}_{\mathbf{Y}_1} & -\underbrace{\mathbf{R}_2 \mathbf{J}_2}_{\mathbf{Y}_2} & \mathbf{0}^{3 \times 3} \\ \mathbf{0}^{3 \times 3} & \mathbf{0}^{3 \times 3} & \underbrace{\mathbf{R}_3 \mathbf{J}_3}_{\mathbf{Y}_3} \\ \underbrace{\mathbf{R}_1 \mathbf{J}_1}_{\mathbf{Y}_1} & \mathbf{W}_1 & \mathbf{W}_2 \end{bmatrix}, \ \mathbf{J}_f^{-1} = \begin{bmatrix} \mathbf{Z}_1 & \mathbf{Z}_2 & \mathbf{Z}_3 \\ \mathbf{Z}_4 & \mathbf{Z}_5 & \mathbf{Z}_6 \\ \mathbf{Z}_7 & \mathbf{Z}_8 & \mathbf{Z}_9 \end{bmatrix} \tag{8.172}$$

$$
\begin{aligned}
\mathbf{Z}_7 &= \mathbf{0}_{3\times3} \\
\mathbf{Z}_8 &= \mathbf{Y}_3^{-1} \\
\mathbf{Z}_9 &= \mathbf{0}_{3\times3} \\
\mathbf{Z}_4 &= -\left(\mathbf{W}_1 + \mathbf{Y}_2\right)^{-1} \\
\mathbf{Z}_1 &= -\mathbf{Y}_1^{-1}\mathbf{W}_1\mathbf{Z}_4 \\
\mathbf{Z}_6 &= -\mathbf{Z}_4 \\
\mathbf{Z}_3 &= \mathbf{Y}_1^{-1}\mathbf{Y}_2\mathbf{Z}_6 \\
\mathbf{Z}_5 &= \mathbf{Z}_4\mathbf{W}_2\mathbf{Z}_8 \\
\mathbf{Z}_2 &= \mathbf{Y}_1^{-1}\mathbf{Y}_2\mathbf{Z}_5.
\end{aligned}
\tag{8.173}
$$

The controlled system was simulated in MATLAB/Simulink. The reference for the center of the triangle formed by the end-effectors of the manipulators was

$$
\mathbf{P}_C^{ref}\left(t\right) = \begin{bmatrix} x_C^{ref}\left(t\right) \\ y_C^{ref}\left(t\right) \\ z_C^{ref}\left(t\right) \end{bmatrix} = \begin{bmatrix} 0.1 + 0.25\sin\left(\pi t\right) \\ 0.25\cos\left(\pi t\right) \\ 1.1 + 0.05\sin\left(\pi t\right) \end{bmatrix}.
\tag{8.174}
$$

The grasping force reference was given as

$$
F_g^{ref}\left(t\right) = 18 + 10\sin\left(\pi t\right)
\tag{8.175}
$$

while the reference for the triangle orientation was

$$
\theta^{ref}\left(t\right) = 0.05 + 0.1\sin\left(\pi t\right).
\tag{8.176}
$$

The initial joint angles of the manipulators were

$$
\mathbf{q}_1\left(0\right) = \begin{bmatrix} 0 \\ -\frac{\pi}{3} \\ \frac{2\pi}{3} \end{bmatrix}, \quad \mathbf{q}_2\left(0\right) = \begin{bmatrix} 0 \\ -\frac{\pi}{3} \\ \frac{2\pi}{3} \end{bmatrix}, \quad \mathbf{q}_3\left(0\right) = \begin{bmatrix} 0 \\ -\frac{\pi}{3} \\ \frac{2\pi}{3} \end{bmatrix}
\tag{8.177}
$$

and the initial velocities were equal to zero. The radius of the circle on which manipulators were located was $P = 2$ m while the radius of the circumscribed circle for which grasping force is equal to zero is $R_0 = 1.05$ m. The properties characterizing the object in the contact points were $K_e = 250$ kg/s^2, $D_e = 5$ kg/s. The parameters used in the control algorithm synthesis were

$$
\begin{aligned}
\mathbf{L}_i &= \operatorname{diag}\left(1500,\ 1500,\ 1500\right),\ i = 1, 2, 3 \\
k_F &= 1000 \\
\mathbf{C} &= \operatorname{diag}\left(c_1, c_2, \ldots, c_9\right),\ c_i = 30,\ i = 1, 2, \ldots, 9 \\
\mathbf{L} &= \operatorname{diag}\left(l_1, l_2, \ldots, l_9\right),\ l_i = 1200,\ i = 1, 2, \ldots, 9 \\
\mathbf{D} &= \operatorname{diag}\left(d_1, d_2, \ldots, d_9\right),\ d_i = 35,\ i = 1, 2, \ldots, 9.
\end{aligned}
\tag{8.178}
$$

The orientation of the triangle formed by the end-effectors was calculated as follows

$$
\begin{aligned}
\mathbf{i} &= \begin{bmatrix} 1 & 0 & 0 \end{bmatrix}^{\mathrm{T}},\ \mathbf{k} = \begin{bmatrix} 0 & 0 & 1 \end{bmatrix}^{\mathrm{T}} \\
\mathbf{c} &= \mathbf{i} \times \left(\mathbf{P_2} - \mathbf{P_1}\right) \\
\theta &= \operatorname{sign}\left(\mathbf{k} \cdot \mathbf{c}\right) \cdot \operatorname{atan2}\left(|\mathbf{c}|, \mathbf{i} \cdot \left(\mathbf{P_2} - \mathbf{P_1}\right)\right).
\end{aligned}
\tag{8.179}
$$

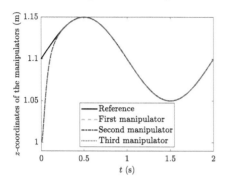

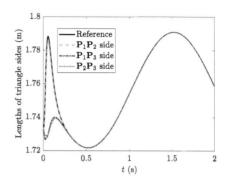

FIGURE 8.32

z-coordinates of the end-effectors.

FIGURE 8.33

Lengths of the triangle sides.

In order to prevent too rapid changes in the length d^{ref}, u_v was bounded so that its minimum and maximum allowed values were -50 and 50 m/s^2. Moreover, to obtain realistic simulation results, all components of the control vector $\mathbf{u}_f$ were also bounded so that their minimum and maximum allowed values were -100 and 100 m/s^2, respectively. Additionally, components of the configuration space input force vectors $\mathbf{T}_1$, $\mathbf{T}_2$, and $\mathbf{T}_3$ were bounded, with -250 and 250 N·m being their minimum and maximum allowed values.

The simulation results are shown in Figures 8.32–8.46. Responses for all coordinates are shown in the global coordinate frame. From Figure 8.32, it is clear that the manipulators' end-effectors are successfully controlled to lie in a plane parallel to the x-y plane. The triangle formed by the end-effectors converges to an equilateral triangle (see Figure 8.33). The triangle center's coordinates and triangle orientation converge to their references, as shown in Figures 8.34–8.37. The grasping force is also successfully controlled, which can be concluded from Figures 8.38 and 8.39. A small undershoot in the grasping force response is due to the high-order dynamics in the force control and saturated control signals. However, the undershoot is small and it can be neglected. The same is valid for small variations (less than 0.035 N) in the force tracking error after the force response converges to its reference. This happens because of the high-order dynamics in the force control. The virtual control u_v reaches its maximum and minimum allowed values at the beginning of the simulation (see Figure 8.40). All components of the vector $\mathbf{u}_f$, except u_{f9} get saturated at the beginning of the simulation (see Figures 8.41–8.43). In the configuration space, only the second joint torque of the second manipulator, T_{22} reaches its limit value for a short time interval at the beginning of the simulation, as depicted in Figures 8.44–8.46.

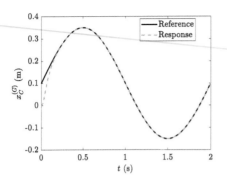

FIGURE 8.34
x-coordinate of the triangle center.

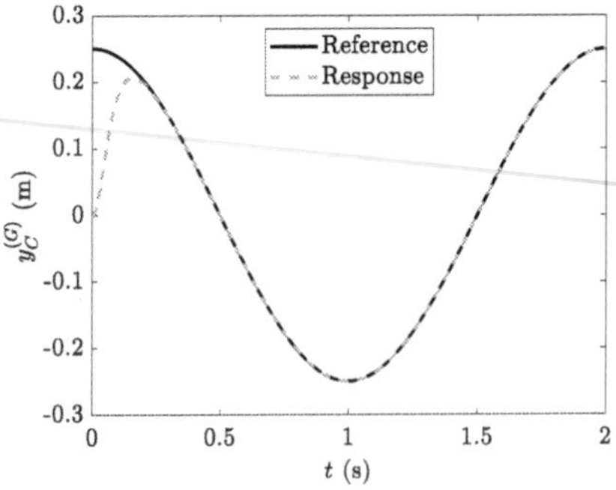

FIGURE 8.35
y-coordinate of the triangle center.

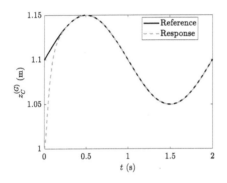

FIGURE 8.36
z-coordinate of the triangle center.

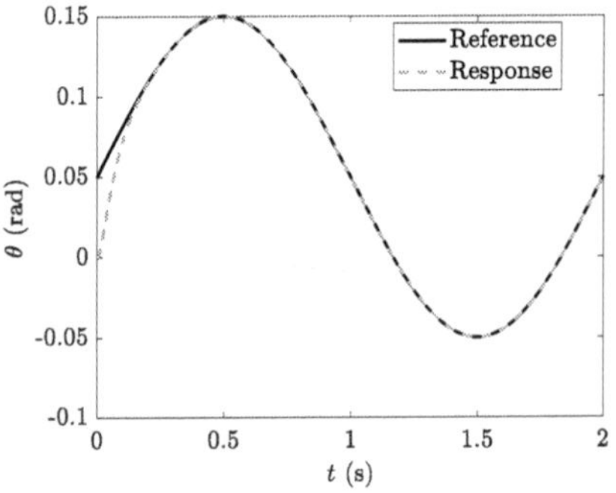

FIGURE 8.37
Orientation of the triangle.

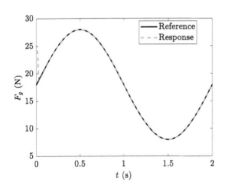

FIGURE 8.38
Grasping force.

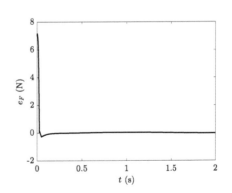

FIGURE 8.39
Grasping force tracking error.

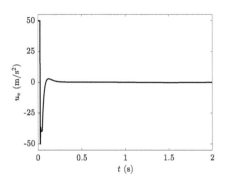

FIGURE 8.40
Virtual control u_v.

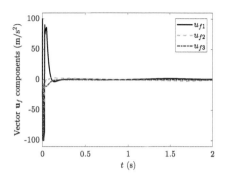

FIGURE 8.41
The first three components of the vector $\mathbf{u}_f$.

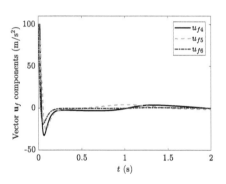

FIGURE 8.42
The second three components of the vector $\mathbf{u}_f$.

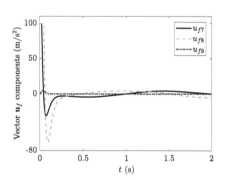

FIGURE 8.43
The last three components of the vector $\mathbf{u}_f$.

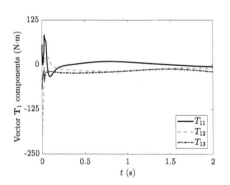

FIGURE 8.44
Vector $\mathbf{T}_1$ components.

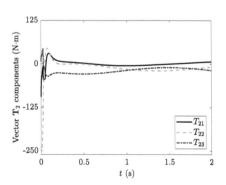

FIGURE 8.45
Vector $\mathbf{T}_2$ components.

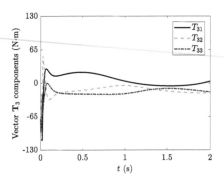

FIGURE 8.46
Vector $\mathbf{T}_3$ components.

8.4   Conclusion

This chapter contains illustrative simulation examples in which the control design approach presented in Chapter 3 has been successfully applied. Two different tasks involving planar manipulators were discussed, the motion synchronization task and object manipulation task. In addition, this chapter deals with control of a bilateral system and control of a system involving three degrees of freedom elbow manipulators doing object manipulation. In all of these examples, the control synthesis was done by the proposed approach. The obtained simulation results showed that all tasks were successfully executed, which demonstrates the effectiveness of the proposed approach for different systems and different tasks.

9

Conclusion

In this book, motion control design for functionally related systems has been discussed. Functionally related systems are systems 'virtually' interconnected through certain functional relations defined by a task that the systems need to execute in a cooperative manner, even though they may be physically separated. The functional relations, which can be expressed as functions of the coordinates of the systems, form a system of 'virtually' interconnected subsystems. Each of these functions represents a functional dependence between the coordinates of the overall system, i.e., coordinates of its subsystems. The method for the task control presented in this book relies on enforcing the functions to track their references.

The presented method is based on representing the dynamics of the controlled system in a new space, called function space. In the configuration space, the control acceleration vector is considered as the control signal, and input force is taken just as a mean to enforce that acceleration through the inverse inertia matrix as control distribution matrix. By representing the task dynamics in the function space using the function Jacobian matrix, and having transformation of the control signals from the function space back to the configuration space by a right pseudoinverse of the function Jacobian matrix, one can achieve that the control distribution matrix in the function space becomes an identity matrix. Therefore, a decoupled control of the functions is possible. In this book, three different control methods were discussed for the control synthesis in the function space: disturbance-observer-based control, sliding mode control, and a method based on the equivalent control estimation. It was shown that the whole control system can have a two-layer structure, and two different forms were shown. In one form, there are low-level controllers that enforce tracking of the references created by a high-level controller and calculated on the basis of controlled functions. This type of the control system was proposed for formation control of mobile robots. In the other form, low-level compensators compensate for disturbances in the configuration space, and a high-level controller now controls the system with compensated disturbances. A control system of this type was proposed for a bilateral system, and also for an object manipulation task in 3-D space. The book has also discussed control of a hierarchical structure of tasks. It was shown that dynamics of lower priority tasks has to be described in the null space of higher priority tasks, so that hierarchy is preserved. The same conclusion can be made if tasks are combined with constraints in the system. In this study, control of redundant tasks has been also elaborated upon. It is demonstrated that a right pseudoinverse of the function Jacobian matrix can be used as an additional degree of freedom in control synthesis. Different pseudoinverse matrices can be used depending on the criterion for selection.

In this work, the selection of functions has been the focus of interest. It has been shown that the rank of the function Jacobian in a system actually shows whether the functions can be controlled in such a way that the desired dynamics is enforced for each function. Moreover, it has been analyzed how selection of the functions and their references can make the controlled system have multiple configurations in the configuration space for which the functions converge to their references. It has been illustrated that the system may end up in any of these configurations, depending on the initial configuration of the system. For a certain selection of the functions and their references, there may be no solution in the configuration space for which the functions converge to their corresponding references.

The proposed approach for control design was applied to several different systems, both in simulations and experiments. It was shown that different motion control problems can be considered in the presented framework and control synthesis can be done in almost identical manner for tasks of different nature, executed by different systems. Thus, different tasks with planar manipulators were executed and obtained simulation and experimental results proved the validity of the presented approach. Moreover, a formation of mobile robots is controlled using the proposed method for control design. In addition, it is illustrated how walking piezoelectric motors can be effectively driven if one is able to identify and control important functions for the motors. Furthermore, object manipulation control in 3-D space is discussed.

Bibliography

[1] K. Abidi and A. Sabanovic, "Sliding-mode control for high-precision motion of a piezostage," *IEEE Transactions on Industrial Electronics*, vol. 54, no. 1, pp. 629–637, 2007.

[2] K. J. Åström and T. Hägglund, *PID Controllers: Theory, Design and Tuning.* Instrument Society of America, Research Triangle Park, NC, USA, 1995.

[3] K. J. Åström and T. Hägglund, "The future of PID control," *Control Engineering Practice*, vol. 9, no. 11, pp. 1163–1175, 2001.

[4] R. Bautista-Quintero and M. J. Pont, "Implementation of H-infinity control algorithms for sensor-constrained mechatronic systems using low-cost microcontrollers," *IEEE Transactions on Industrial Informatics*, vol. 4, no. 3, pp. 175–184, 2008.

[5] K.-W. Byun, B. Wie, D. Geller, and J. Sunkel, "Robust H_∞ control design for the space station with structured parameter uncertainty," *Journal of Guidance, Control, and Dynamics*, vol. 14, no. 6, pp. 1115–1122, 1991.

[6] M. Canale, L. Fagiano, A. Ferrara, and C. Vecchio, "Vehicle yaw control via second-order sliding-mode technique," *IEEE Transactions on Industrial Electronics*, vol. 55, no. 11, pp. 3908–3916, 2008.

[7] I. Cervantes and J. Alvarez-Ramirez, "On the PID tracking control of robot manipulators," *Systems & Control Letters*, vol. 42, no. 1, pp. 37–46, 2001.

[8] G. Chen and T. T. Pham, *Introduction to fuzzy sets, fuzzy logic, and fuzzy control systems.* CRC Press, Boca Raton, FL, USA, 2000.

[9] J. Cheng, J. Yi, and D. Zhao, "Design of a sliding mode controller for trajectory tracking problem of marine vessels," *IET Control Theory & Applications*, vol. 1, no. 1, pp. 233–237, 2007.

[10] S. Chiu, S. Chand, D. Moore, and A. Chaudhary, "Fuzzy logic for control of roll and moment for a flexible wing aircraft," *IEEE Control Systems*, vol. 11, no. 4, pp. 42–48, 1991.

[11] S. Cong and Y. Liang, "PID-like neural network nonlinear adaptive control for uncertain multivariable motion control systems," *IEEE Transactions on Industrial Electronics*, vol. 56, no. 10, pp. 3872–3879, 2009.

[12] D. L. DeVoe and A. P. Pisano, "Modeling and optimal design of piezoelectric cantilever microactuators," *Journal of Microelectromechanical Systems*, vol. 6, no. 3, pp. 266–270, 1997.

[13] B. Draženović, "The invariance conditions in variable structure systems," *Automatica*, vol. 5, no. 3, pp. 287–295, 1969.

[14] S. Emelyanov, *Theory of Systems with Variable Structure*.   Nauka, Moscow, Russia, 1970 (in Russian).

[15] E. D. Engeberg, S. G. Meek, and M. A. Minor, "Hybrid force–velocity sliding mode control of a prosthetic hand," *IEEE Transactions on Biomedical Engineering*, vol. 55, no. 5, pp. 1572–1581, 2008.

[16] B. A. Francis, *A Course in H_∞ Control Theory*.   Springer-Verlag, New York, USA, 1987.

[17] G. F. Franklin, J. D. Powell, and A. Emami-Naeini, *Feedback Control of Dynamic Systems*.   Prentice Hall, Upper Saddle River, NJ, USA, 2001.

[18] E. Golubovic, E. A. Baran, and A. Sabanovic, "Contouring controller for precise motion control systems," *Automatika*, vol. 54, no. 1, 2013.

[19] E. Golubovic, T. Uzunovic, Z. Zhakypov, and A. Sabanovic, "Adaptive control of piezoelectric walker actuator," in *Proceedings of the IEEE International Conference on Mechatronics*, 2013, pp. 132–137.

[20] E. Golubovic, Z. Zhakypov, T. Uzunovic, and A. Sabanovic, "Piezoelectric motor driver: Design and evaluation," in *Proceedings of the IECON 2013-39th Annual Conference of the IEEE Industrial Electronics Society*, 2013, pp. 3964–3969.

[21] J. Guldner and V. I. Utkin, "Sliding mode control for gradient tracking and robot navigation using artificial potential fields," *IEEE Transactions on Robotics and Automation*, vol. 11, no. 2, pp. 247–254, 1995.

[22] A. Hace, K. Jezernik, and A. Sabanovic, "SMC with disturbance observer for a linear belt drive," *IEEE Transactions on Industrial Electronics*, vol. 54, no. 6, pp. 3402–3412, 2007.

[23] F. Harashima, J. X. Xu, and H. Hashimoto, "Tracking control of robot manipulators using sliding mode," *IEEE Transactions on Power Electronics*, vol. PE-2, no. 2, pp. 169–176, 1987.

[24] T. Hiraoka, O. Nishihara, and H. Kumamoto, "Model-following sliding mode control for active four-wheel steering vehicle," *Review of Automotive Engineering*, vol. 25, no. 3, p. 305, 2004.

[25] G. W. Irwin, K. Warwick, and K. J. Hunt (Eds.), *Neural Network Applications in Control.* The Institution of Electrical Engineers, London, UK, 1995.

[26] R. Isermann, *Mechatronic Systems: Fundamentals.* Springer-Verlag, London, UK, 2005.

[27] S. Ishikawa, "A method of indoor mobile robot navigation by using fuzzy control," in *Proceedings of the IEEE/RSJ International Workshop on Intelligent Robots and Systems*, 1991, pp. 1013–1018.

[28] U. Itkis, *Control Systems of Variable Structure.* Wiley, New York, USA, 1976.

[29] S. Johansson, M. Bexell, and A. Jansson, "Fine control of electromechanical motors," U.S. Patent 6 798 117, Sep. 28, 2004.

[30] S. Johansson, M. Bexell, and P. O. Lithell, "Fine walking actuator," U.S. Patent 6 337 532, Jan. 8, 2002.

[31] S. Johansson, M. Bexell, and P. O. Lithell, "Switched actuator control," U.S. Patent 6 459 190, Oct. 1, 2002.

[32] M. S. Ju, C. C. Lin, D. H. Lin, I. S. Hwang, and S. M. Chen, "A rehabilitation robot with force-position hybrid fuzzy controller: hybrid fuzzy control of rehabilitation robot," *IEEE Transactions on Neural Systems and Rehabilitation Engineering*, vol. 13, no. 3, pp. 349–358, 2005.

[33] S. Katsura, W. Iida, and K. Ohnishi, "Medical mechatronics – an application to haptic forceps," *Annual Reviews in Control*, vol. 29, no. 2, pp. 237–245, 2005.

[34] S. Katsura and K. Ohnishi, "Human cooperative wheelchair for haptic interaction based on dual compliance control," *IEEE Transactions on Industrial Electronics*, vol. 51, no. 1, pp. 221–228, 2004.

[35] S. Kawamura, F. Miyazaki, and S. Arimoto, "Is a local linear PD feedback control law effective for trajectory tracking of robot motion?" in *Proceedings of the IEEE International Conference on Robotics and Automation*, vol. 3, 1988, pp. 1335–1340.

[36] O. Khatib, "A unified approach for motion and force control of robot manipulators: The operational space formulation," *IEEE Journal of Robotics and Automation*, vol. 3, no. 1, pp. 43–53, 1987.

[37] Y. H. Kim and F. L. Lewis, "Neural network output feedback control of robot manipulators," *IEEE Transactions on Robotics and Automation*, vol. 15, no. 2, pp. 301–309, 1999.

[38] S. Komada, M. Ishida, K. Ohnishi, and T. Hori, "Disturbance observer-based motion control of direct drive motors," *IEEE Transactions on Energy Conversion*, vol. 6, no. 3, pp. 553–559, 1991.

[39] C. Y. Lai, F. L. Lewis, V. Venkataramanan, X. Ren, S. S. Ge, and T. Liew, "Disturbance and friction compensations in hard disk drives using neural networks," *IEEE Transactions on Industrial Electronics*, vol. 57, no. 2, pp. 784–792, 2010.

[40] S. Lin and A. A. Goldenberg, "Neural-network control of mobile manipulators," *IEEE Transactions on Neural Networks*, vol. 12, no. 5, pp. 1121–1133, 2001.

[41] C.-C. Lo and C.-Y. Chung, "Tangential-contouring controller for biaxial motion control," *Journal of Dynamic Systems, Measurement, and Control*, vol. 121, no. 1, pp. 126–129, 1999.

[42] R. J. E. Merry, N. C. T. de Kleijn, M. J. G. van de Molengraft, and M. Steinbuch, "Using a walking piezo actuator to drive and control a high-precision stage," *IEEE/ASME Transactions on Mechatronics*, vol. 14, no. 1, pp. 21–31, 2009.

[43] R. J. E. Merry, M. G. J. M. Maassen, M. J. G. van de Molengraft, N. van de Wouw, and M. Steinbuch, "Modeling and waveform optimization of a nano-motion piezo stage," *IEEE/ASME Transactions on Mechatronics*, vol. 16, no. 4, pp. 615–626, 2011.

[44] T. Murakami and K. Ohnishi, "Observer-based motion control-application to robust control and parameter identification," in *Proceedings of the Asia-Pacific Workshop on Advances in Motion Control*, 1993, pp. 1–6.

[45] T. Murakami, F. Yu, and K. Ohnishi, "Torque sensorless control in multidegree-of-freedom manipulator," *IEEE Transactions on Industrial Electronics*, vol. 40, no. 2, pp. 259–265, 1993.

[46] R. M. Murray, Z. Li, and S. S. Sastry, *A Mathematical Introduction to Robotic Manipulation*. CRC press, Boca Raton, FL, USA, 1994.

[47] K. Ohishi, K. Ohnishi, and K. Miyachi, "Torque-speed regulation of DC motor based on load torque estimation," in *Proceedings of the IEEJ International Power Electronics Conference*, vol. 2, 1983, pp. 1209–1216.

[48] K. Ohnishi, M. Shibata, and T. Murakami, "Motion control for advanced mechatronics," *IEEE/ASME Transactions on Mechatronics*, vol. 1, no. 1, pp. 56–67, 1996.

[49] J. Park and W. K. Chung, "Analytic nonlinear H_∞ inverse-optimal control for euler-lagrange system," *IEEE Transactions on Robotics and Automation*, vol. 16, no. 6, pp. 847–854, 2000.

[50] PiezoMotor Uppsala AB. (2014, Nov.) Linear Piezo LEGS LL1011A motor datasheet. [Online]. Available: http://www.piezomotor.com/app/content/uploads/150010_LL10.pdf

[51] P. Rocco, "Stability of PID control for industrial robot arms," *IEEE Transactions on Robotics and Automation*, vol. 12, no. 4, pp. 606–614, 1996.

[52] A. Rojko and K. Jezernik, "Sliding-mode motion controller with adaptive fuzzy disturbance estimation," *IEEE Transactions on Industrial Electronics*, vol. 51, no. 5, pp. 963–971, 2004.

[53] A. Šabanović, K. Jezernik, and K. Wada, "Chattering-free sliding modes in robotic manipulators control," *Robotica*, vol. 14, no. 01, pp. 17–29, 1996.

[54] A. Šabanović and K. Ohnishi, *Motion Control Systems*. John Wiley & Sons, Singapore, 2011.

[55] C. O. Saglam, E. A. Baran, A. O. Nergiz, and A. Sabanovic, "Model following control with discrete time smc for time-delayed bilateral control systems," in *Proceedings of the IEEE International Conference on Mechatronics*, 2011, pp. 997–1002.

[56] S. Sakaino, T. Sato, and K. Ohnishi, "Oblique coordinate control for advanced motion control-applied to micro-macro bilateral control," in *Proceedings of the 2009 IEEE International Conference on Mechatronics*, 2009, pp. 1–6.

[57] V. Sankaranarayanan and A. D. Mahindrakar, "Control of a class of underactuated mechanical systems using sliding modes," *IEEE Transactions on Robotics*, vol. 25, no. 2, pp. 459–467, 2009.

[58] T. Shibata and T. Murakami, "Null space motion control by PID control considering passivity in redundant manipulator," *IEEE Transactions on Industrial Informatics*, vol. 4, no. 4, pp. 261–270, 2008.

[59] M. W. Spong and M. Vidyasagar, *Robot Dynamics and Control*. John Wiley & Sons, USA, 1989.

[60] C. Y. Su and Y. Stepanenko, "Adaptive variable structure tracking control for constrained robots," *IEEE Transactions on Aerospace and Electronic Systems*, vol. 30, no. 2, pp. 493–503, 1994.

[61] Y. L. Sun and M. J. Er, "Hybrid fuzzy control of robotics systems," *IEEE Transactions on Fuzzy Systems*, vol. 12, no. 6, pp. 755–765, 2004.

[62] F. Szufnarowski and A. Schneider, "Two-dimensional dynamics of a quasi-static legged piezoelectric actuator," *Smart Materials and Structures*, vol. 21, no. 5, 2012.

[63] H. G. Tanner, G. J. Pappas, and V. Kumar, "Leader-to-formation stability," *IEEE Transactions on Robotics and Automation*, vol. 20, no. 3, pp. 443–455, 2004.

[64] S. Timoshenko *et al.*, "Analysis of bi-metal thermostats," *Journal of the Optical Society of America*, vol. 11, no. 3, pp. 233–255, 1925.

[65] G. J. Toussaint, T. Basar, and F. Bullo, "H^∞-optimal tracking control techniques for nonlinear underactuated systems," in *Proceedings of the 39th IEEE Conference on Decision and Control*, vol. 3, 2000, pp. 2078–2083.

[66] T. Tsuji, K. Ohnishi, and A. Sabanovic, "A controller design method based on functionality," in *Proceedings of the 9th IEEE International Workshop on Advanced Motion Control*, 2006, pp. 171–176.

[67] T. Tsuji, K. Ohnishi, and A. Sabanovic, "A controller design method based on functionality," *IEEE Transactions on Industrial Electronics*, vol. 54, no. 6, pp. 3335–3343, 2007.

[68] V. Utkin, J. Guldner, and J. Shi, *Sliding Mode Control in Electromechanical Systems*. CRC Press, Boca Raton, FL, USA, 2009.

[69] V. I. Utkin, S. V. Drakunov, H. Hashimoto, and F. Harashima, "Robot path obstacle avoidance control via sliding mode approach," in *Proceedings of the IEEE/RSJ International Workshop on Intelligent Robots and Systems*, 1991, pp. 1287–1290.

[70] V. I. Utkin, *Sliding Modes and Their Application in Variable Structure Systems*. Mir Publishers, Moscow, Russia, 1978.

[71] V. I. Utkin, *Sliding Modes in Control and Optimization*. Springer-Verlag, Berlin, Germany, 1992.

[72] T. Uzunovic, "Motion control design for functionally related systems," Ph.D. dissertation, Sabanci University, Istanbul, Turkey, Aug. 2015.

[73] T. Uzunovic, E. Golubovic, E. A. Baran, and A. Sabanovic, "Configuration space control of a parallel delta robot with a neural network based inverse kinematics," in *Proceedings of the 8th International Conference on Electrical and Electronics Engineering*, 2013, pp. 497–501.

[74] T. Uzunovic, E. Golubovic, and A. Sabanovic, "FPGA based control of a walking piezo motor," in *Proceedings of the IEEE 13th International Workshop on Advanced Motion Control*, 2014, pp. 138–143.

[75] T. Uzunovic, E. Golubovic, and A. Sabanovic, "Piezo LEGS driving principle based on coordinate transformation," *IEEE/ASME Transactions on Mechatronics*, vol. 20, no. 3, pp. 1395–1405, 2015.

[76] T. Uzunovic, E. Golubovic, and A. Sabanovic, "Force control of piezoelectric walker," in *Proceedings of the IECON 2016-42nd Annual Conference of the IEEE Industrial Electronics Society*, 2016, pp. 5790–5795.

[77] T. Uzunovic and A. Sabanovic, "Formation control of differential-drive mobile robots in the framework of functionally related systems," in *Proceedings of the IECON 2015-41st Annual Conference of the IEEE Industrial Electronics Society*, 2015, pp. 002 620–002 625.

[78] T. Uzunovic and A. Sabanovic, "A novel approach to motion control design for functionally related systems," *International Journal of Control, Automation and Systems*, vol. 16, pp. 1–12, 2018.

[79] T. Uzunovic, J. Velagic, N. Osmic, A. Badnjevic, and E. Zunic, "Neural networks for helicopter azimuth and elevation angles control obtained by cloning processes," in *Proceedings of the IEEE International Conference on Systems Man and Cybernetics*, 2010, pp. 1076–1082.

[80] A. J. van der Schaft, "L_2-gain analysis of nonlinear systems and nonlinear state-feedback H_∞ control," *IEEE Transactions on Automatic Control*, vol. 37, no. 6, pp. 770–784, 1992.

[81] A. Šabanović, "Variable structure systems with sliding modes in motion control – a survey," *IEEE Transactions on Industrial Informatics*, vol. 7, no. 2, pp. 212–223, 2011.

[82] S. Weerasooriya and M. A. El-Sharkawi, "Identification and control of a dc motor using back-propagation neural networks," *IEEE Transactions on Energy Conversion*, vol. 6, no. 4, pp. 663–669, 1991.

[83] A. Wege, K. Kondak, and G. Hommel, "Force control strategy for a hand exoskeleton based on sliding mode position control," in *Proceedings of the IEEE/RSJ International Conference on Intelligent Robots and Systems*, 2006, pp. 4615–4620.

[84] K. A. Wise and J. L. Sedwick, "Nonlinear H_∞ optimal control for agile missiles," *Journal of Guidance, Control, and Dynamics*, vol. 19, no. 1, pp. 157–165, 1996.

[85] B. Yao, S. P. Chan, and D. Wang, "Variable structure adaptive motion and force control of robot manipulators," *Automatica*, vol. 30, no. 9, pp. 1473–1477, 1994.

[86] Y. Yokokohji and T. Yoshikawa, "Bilateral control of master-slave manipulators for ideal kinesthetic coupling–formulation and experiment," *IEEE Transactions on Robotics and Automation*, vol. 10, no. 5, pp. 605–620, 1994.

[87] B. K. Yoo and W. C. Ham, "Adaptive control of robot manipulator using fuzzy compensator," *IEEE Transactions on Fuzzy Systems*, vol. 8, no. 2, pp. 186–199, 2000.

[88] K.-K. D. Young, "A variable structure model following control design for robotics applications," *IEEE Journal of Robotics and Automation*, vol. 4, no. 5, pp. 556–561, 1988.

[89] K. K. Young, "Controller design for a manipulator using theory of variable structure systems," *IEEE Transactions on Systems, Man and Cybernetics*, vol. 8, no. 2, pp. 101–109, 1978.

[90] L. A. Zadeh, "Fuzzy sets," *Information and Control*, vol. 8, no. 3, pp. 338–353, 1965.

[91] L. A. Zadeh, "Outline of a new approach to the analysis of complex systems and decision processes," *IEEE Transactions on Systems, Man and Cybernetics*, vol. SMC-3, no. 1, pp. 28–44, 1973.

[92] G. Zhang, "Speed control of two-inertia system by PI/PID control," *IEEE Transactions on Industrial Electronics*, vol. 47, no. 3, pp. 603–609, 2000.

Index